Future Competition in Telecommunications

Future
Competition in
Telecommunications

Edited by
Stephen P. Bradley
and Jerry A. Hausman

HARVARD BUSINESS SCHOOL PRESS
Boston, Massachusetts

93 92 91 90 89   5 4 3

Library of Congress Catologing-in-Publication Data

Future competition in telecommunications.

   Includes bibliographies and index.
   1. Telecommunication—United States.   2. Telecommunication—United States—Deregulation.
3. Competition—United States.   I. Bradley, Stephen P.,
1941–   . II. Hausman, Jerry A.
HE7775.F88   1989   384'.041   88-35787
ISBN 0-87584-211-9

Contents

Preface

The purpose of this book is to provide insight into the future competitive structure of the telecommunications industry in the United States. Much has been written about the history of the industry and particularly about the main architect of this history, AT&T. Even more has been presented by journalists chronicling the moves and countermoves of the various competitors as they aggressively pushed for or attempted to block deregulation while simultaneously positioning themselves for its ultimate coming. A thorough investigation of this complicated history will eventually reveal a great many lessons; however, rather than attempt to uncover these lessons while they are still unfolding, in this book we attempt to understand what is possible for the future of the telecommunications industry and in some small way help to shape that future.

The situation in the industry in the United States today is one of evolving deregulation. No one has a clear understanding of how the industry will ultimately be deregulated and a time table by which to make the necessary strategic preparations. The current and future competitors confront an environment whose structure is changing in an experimental fashion, the short-term effects of relatively minor experimentation helping to shape the next stage in its evolutionary path.

The basic blueprint for the deregulation of the telecommunications industry in the United States is the Modified Final Judgment (MFJ) executed by Judge Harold Greene of the Federal District Court of the District of Columbia. The MFJ lays out the framework for deregulation of the industry in which AT&T agreed to divest its Bell Operating Companies (BOCs), forming seven independent regional holding companies to provide local telecommunications services. In exchange, AT&T was to become substantially less regulated and be able to enter important new business areas, including computers and certain information services. The BOCs were specifically prohibited from providing long-distance service, manufacturing or providing telecommunications equipment, or entering information services markets.

At the time, the Federal Communications Commission (FCC), whose charter is to provide regulatory oversight of all telecommunications from the federal government's perspective, was on record as disagreeing with many of the MFJ's restrictions on the BOCs. The BOCs have subsequently been permitted to provide, but not manufacture, telecommunications equipment; and recently Judge Greene removed many of the restrictions placed on their providing information services. On the other hand, the FCC designated AT&T a Dominant Carrier so that AT&T's freedom from regulation has been considerably less than it expected. Only lately has the FCC permitted AT&T more pricing leeway for its large users. The price cap proceedings, which are ongoing at the FCC, could provide considerably more regulatory flexibility for AT&T. The result is that the MFJ does not determine in any concrete sense the future structure of the telecommunications industry, since it is subject to review by the Department of Justice and the Decree Court every three years to see if its terms should be altered; and the industry is also subject to evolving federal regulation, through the FCC, and state regulation.

In order to help understand and influence the future structure of the nation's telecommunications industry, a colloquium was held at the Harvard Business School in May 1987. To a certain extent, the rulings of Judge Greene in the fall of 1987 and the spring of 1988 reflect the discussions that took place during the colloquium. But more important, the purpose of the colloquium was to bring together representatives of the various factions so that all participants in the ongoing dialogue would better understand one another from a business perspective—the goal being that this improved understanding would have a lasting impact on the evolution of the industry.

The participants in the colloquium were from five groups: 1) companies that provide telecommunications services and products, including AT&T, the regional holding companies, independent telephone companies, suppliers of long-distance services, regional bypass companies, equipment suppliers, and information services providers; 2) companies that are substantial users of telecommunications services, including financial institutions, manufacturing companies, and service companies; 3) officials involved in the regulation of the industry, including members of the Department of Justice, the FCC, the National Telecommunications and Information Administration (NTIA) of the Department of Commerce, and state regulatory bodies; 4) lawyers who represent each of the above groups; and 5) individuals who study the industry, including academics and consultants. It is hoped that by bringing this diverse group of individuals together to debate the future of the industry a dialogue has been initiated that will continue and help to sensibly frame the heated discussions that will inevitably take place over the next several years. We trust that this book is a useful step in this ongoing process.

It is of course impossible to thank all of the contributors to this effort individually. However, we would like to thank all the participants in the colloquium collectively for spending two and a half days in Boston and creating a very lively discussion of future competition in telecommunications. The authors of the ten chapters in this book played a major role in the success of the colloquium. However, what makes the book more than a collection of papers presented at a colloquium is that each topic addressed by a chapter was discussed and debated at length during the colloquium and the chapters were then substantially revised to reflect the debate that took place.

We would specifically like to thank those participants who were formal discussants during the various sessions. These include H. Brian Thompson, executive vice president, MCI; Dr. Thomas Spavins, deputy chief, Office of Plans and Policy, FCC; Arch McGill, president, Chardonnay, Inc. (formerly president, AT&T Advanced Information Systems); Edward Goldstein, principal, Management Analysis Center, Inc.; John D. Ziglis, senior vice president and general counsel, AT&T; Howard Anderson, president, The Yankee Group; Charles L. Jackson, president, Shooshan & Jackson; Richard McCormick, president and chief operating officer, USWest, Inc.; Gerald Ely, vice president, Merrill Lynch Pierce Fenner and Smith, Inc.; and Paul Levy, chairman, Massachusetts Department of Public Utilities. In addition, we would like to thank those who acted as moderators of specific sessions: Professors Elon Kohlberg and Richard H.K. Vietor, Harvard Business School; and Professor Richard

Schmalensee, Sloan School of Management, Massachusetts Institute of Technology.

We would also like to especially thank two individuals who played very important roles in making the colloquium a success. The first is Professor William Baxter, Stanford Law School, who was assistant attorney general for antitrust when the Consent Decree was negotiated. He played an important role in developing and negotiating the ultimate structure of the breakup of AT&T and he provided a very candid discussion of those events as a way of putting the current set of key issues facing the industry in perspective. Second, we would like to thank William McGowen, president and chief executive officer of MCI, for his enthusiastic encouragement of our efforts and his help in identifying appropriate individuals to include in the colloquium. We only regret that because of health reasons he was unable to attend the sessions.

The Harvard Business School sponsored the colloquium and provided substantial support to make the endeavor a success. We would like to thank Dean John McArthur for inviting Jerry Hausman to visit the Harvard Business School for the academic year 1986–1987, thus providing the opportunity for collaboration, and Senior Associate Dean Jay Lorsch for putting the resources of the Division of Research of the School behind the project.

People behind the scenes always make very important contributions in projects of this magnitude. We owe a debt of gratitude to Rowena Foss, Martha Laisne, and Bridget Testa for providing crucial administrative assistance in planning, executing, and following-up on the colloquium. Finally, we would like to thank Barbara Ankeny and Natalie Greenberg of the Harvard Business School Press for their encouragement and follow-through, respectively, in developing and editing this book.

Boston, Massachusetts   Stephen P. Bradley
Cambridge, Massachusetts   Jerry A. Hausman
October 1988

Future Competition in Telecommunications

1
Future Competition in Telecommunications

Stephen P. Bradley and Jerry A. Hausman

On January 1, 1984, the United States undertook a new policy in telecommunications. On that day the divestiture of AT&T took place according to the Modified Final Judgment (MFJ). AT&T was divided into eight corporate parts. Seven parts were regional holding companies, which included the Bell operating companies (BOCs) that were permitted to provide local telecommunications services. The remaining part was AT&T, which remained by far the largest supplier of long-distance telecommunications service and of many types of telecommunications equipment. This new policy was not enacted by Congress; indeed, the United States still labors under the Communications Act of 1934, which sets the framework for the federal government's involvement in telecommunications. Nor did the Federal Communications Commission (FCC) put the policy into place. The FCC was on record as disagreeing with many of the restrictions placed on the BOCs by the MFJ.[1] The transformation of U.S. telecommunications took place under an antitrust consent decree between AT&T and the U.S. Department of Justice. This consent decree, which was substantially modified from its original form, has been executed by a single federal judge, Judge Harold Greene of the Federal District Court of the District of Columbia.

This extraordinary situation provided the background of a colloquium, "Future Competition in Telecommunications," that took place at the Harvard Business School on May 28–29, 1987.

Note: Paul White and Roger Noll provided helpful comments on this chapter, although Professor Noll disagrees with much of the analysis.

The purpose of the colloquium was to bring together those representatives of the major participants in the debate who would undoubtedly play important roles in determining the future competitive structure of the industry. The colloquium was organized around five groups: businesses that provide telecommunications products and services; businesses that are substantial users of telecommunications; FCC officials who regulate telecommunications and Department of Justice officials who enforce the MFJ; state regulatory officials; and academics and consultants who study telecommunications. Nine papers were commissioned to provide the framework for discussions of *future* competition in telecommunications. The appropriate background for the colloquium was established with an introduction by William Baxter, who was assistant attorney general in charge of antitrust when the consent decree was negotiated, and a paper by Richard H.K. Vietor, "AT&T and the Public Good: Regulation and Competition in Telecommunications, 1910–1987," which is Chapter 2 of this volume; both discussed the regulatory history and economic framework of the MFJ.

We specifically requested that conference participants refrain, at least during the conference sessions, from relitigation of old cases or discussion of past issues, such as whether predivestiture AT&T had used long-distance revenues to cross-subsidize local service, as most individuals who have studied the matter conclude, or whether AT&T had used local revenues to cross-subsidize long-distance service to hinder competition. Instead, we attempted to concentrate attention on the future evolution of telecommunications in the United States within the framework of the MFJ and evolving federal (FCC) and state regulation.

As the conference took place, the future framework for competition in telecommunications was itself particularly unsettled. Under the terms of the MFJ, a review by the Department of Justice and the decree court takes place every three years to ascertain whether terms of the MFJ should be altered. This first triennial review began in February 1987, when the Department of Justice published its own report, *The Geodesic Network: 1987 Report on Competition in the Telephone Industry* (U.S. Department of Justice, 1987), written by consultant Peter Huber, and released its recommendations to Judge Greene for modification of the MFJ. In Chapter 3, "The Technological Imperative for Competition," Huber discusses his view that telecommunications is largely technology driven and the effect that future technological advances will have on competition in telecommunications. The Department of Justice called for removal of many of the restrictions placed on the BOCs by the MFJ.[2] The likely effect of the Justice Department's recommendations would have been to increase competition in certain telecommunications markets and to encourage the development of telecommunications services, especially for residential customers, not currently being provided in an efficient manner in the United States.

The current and potential future competitors of the BOCs, as well as some customer groups, did not see the situation in a similar manner. More than 170 parties filed comments with the court, either supporting or contradicting Huber's analysis and the Justice Department's recommendations.[3] At the time of the conference essentially all these comments had been filed with the court. Most of the conference participants awaited Judge Greene's ruling with great interest. When Judge Greene made his ruling in September 1987, he removed some restrictions that the MFJ had imposed on permissible BOC activities. Most important, in his September 1987 decision and in a subsequent ruling in March 1988, Judge Greene removed many of the restrictions on BOC provision of information services. However, he rejected the majority of the other recommendations proposed by the Department of Justice with the support of both the FCC and the National Telecommunications and Information Administration (NTIA) of the Department of Commerce, the other two federal agencies engaged in setting telecommunications policy.

Thus, U.S. telecommunications policy is currently in an exceedingly peculiar situation.[4] The three federal agencies that have an oversight role on telecommunications have a very wide area of agreement that the MFJ restrictions on the BOCs should be relaxed significantly. However, the federal district judge who issued the MFJ seemingly has the ability to overrule them within his interpretation of the relevant antitrust law. And Congress, which has the ability to resolve this jurisdictional stalemate by legislation, appears extremely unlikely to act in the face of the conflicting business and economic interests involved in the debate over the future competitive framework for telecommunications.[5]

Meanwhile, the information age continues to evolve with the rapid globalization of financial markets, increased dependence of manufacturing on telecommunications, and advances in semiconductor (microchip) technology that will have important effects on both business and residential telecommunications demands. The combination of technology and economics may in time make the current debate one of only historical interest. However, given the present competitive policy toward telecommunications, the maximum benefits from competition will be achieved when U.S. policy is made in the framework of the technological and economic realities of today, not the now outdated framework of the 1970s and the predivestiture AT&T period when both the Justice Department and the FCC struggled unsuccessfully to control AT&T's economic and political power.

The Competitive Framework of the MFJ

The MFJ divided the world of telecommunications services into two parts. The boundary between the monopoly sector controlled by the BOCs and the competitive sector open to anyone but the BOCs was set by the Local Access and Transport Areas (LATAs), which correspond approximately to government-defined Consolidated Metropolitan Statistical Areas, although in some cases entire states constitute a single LATA. Three main telecommunications services are provided by the BOCs within their LATA boundaries: local exchange service (local calls); exchange access service, which is the equal access service provided by the BOCs for long-distance (interexchange) carriers; and intraLATA long-distance, which may be as short as a call from Wall Street to Brooklyn or as long as a call from the eastern end of Long Island to White Plains. The MFJ prohibits the BOCs from participation in the interexchange, interLATA service markets in which AT&T, MCI, US Sprint, and other companies provide both voice and data telecommunications services; Telenet, Tymnet, and other companies provide essentially data telecommunications services. The commonly held expectation at the time the MFJ was issued in 1982 was that the BOCs held a "bottleneck monopoly," which would make competition extremely unlikely, if not impossible, within the LATA boundaries. However, the technological and economic framework was expected to permit competition for services that crossed LATA boundaries.[6]

BOC participation in the competitive sector was forbidden because the BOCs were judged to have the "incentives and abilities" to impede competition in long-distance (interLATA) markets. The incentive supposedly arose from state regulation that limits BOC prices and profits from provision of local services. The economic theory used to justify the restrictions on the BOCs then holds that the BOCs will have the incentive to transfer profits from the regulated sector to the competitive interexchange markets, where regulatory restraints will be absent or at least much less constraining.[7]

The ability to impede competition was held to arise from the BOC bottleneck monopoly over exchange access. All interexchange competitors were expected to be dependent on BOC-provided equal access to their networks. If the BOCs were allowed to compete in interexchange networks, they could discriminate against their competitors in either service quality or price.[8] Thus, so long as the bottleneck monopoly remained, it was held that the BOCs could leverage their monopoly to impede competition in interexchange markets, so that their participation was to be prohibited.

The "incentives and abilities" theory for interexchange services was also applied in the MFJ to BOC provision of information services and the manufacture of telecommunications equipment. The "line of business" restrictions in Section II(D) of the MFJ prohibit the following:

> [N]o BOC shall, directly or through any affiliated enterprise:
> 1. Provide interexchange telecommunications services or information services;
> 2. Manufacture or provide telecommunications products or customer premises equipment (except for provision of customer premises equipment for emergency services); or
> 3. Provide any other product or service, except exchange telecommunications and exchange access service, that is not a natural monopoly service actually regulated by tariff.[9]

Under Section VIII(C) of the MFJ, the II(D) restrictions will be eliminated if "there is no substantial possibility that it [the BOC] could use its monopoly power to impede competition in the market it seeks to enter." Thus, under the terms of the MFJ, the BOCs were forbidden to enter *any* enterprise except provision of local telecommunications service without the permission of the federal district court.

The BOCs were required to receive waivers from the court before they could enter nontelecommunications markets or provide telecommunications services now prohibited by Section II(D) of the MFJ. More than 150 of these waivers have been granted, with BOCs now engaged in the provision of insurance, real estate services, equipment distribution, and financial services, as well as cellular mobile radio (car telephone) service outside their regions.

The Justice Department's recommendations to Judge Greene in the first triennial review of the MFJ called for elimination of Sections II(D)2 and 3 so that the BOCs could enter into manufacturing markets and nontelecommunications markets without the need of a waiver. The Justice Department also recommended that Section II(D)1 be revised so that the BOCs could provide information services. However, the interexchange prohibition of Section II(D)1 was to continue.

In his September 1987 opinion, Judge Greene declined to follow the recommendations of the U.S. government. He did remove the nontelecommunications restriction, II(D)3. He also said that he intended to relax the information service restriction, II(D)1, by allowing the BOCs to provide information service transmission and gateway facilities. Judge Greene would not allow the BOCs to provide information service content, such as electronic Yellow Pages or the databases that permit computer processing and response to information requests. In March 1988, Judge Greene made final his decision to allow BOC provision of transmission and gateway information services. He also permitted the BOCs to provide other information services such as voice messaging, that is, electronic answering and calling services. The Section II(D)2 manufacturing restriction on the BOCs as well as the interexchange restriction of Section II(D)1 remained in place.

Potentially at least as important as his rulings on the specific government recommendations, Judge Greene ruled that given its past lack of success in regulation of predivestiture AT&T, the FCC could not be expected to regulate effectively the postdivestiture BOCs. Given the mistrust of potential regulatory solutions to the "incentives and abilities" problem that the government had advanced, the so-called core restrictions on interexchange service and on manufacturing seem likely to remain in place for at least another three years unless Congress acts or unless an appeal to a higher court by a BOC or by the Justice Department is successful. The legal and regulatory situation currently in place is analyzed by Kevin R. Sullivan in Chapter 4, "Competition in Telecommunications: Moving Toward a New Era of Antitrust Scrutiny and Regulation." Sullivan sees the possibility of a retreat from the increased competition in telecommunications of the past decade as individual states exert their authority to regulate telecommunications under the "state action doctrine."

Important regulatory changes have taken place since the AT&T divestiture. These changes are likely to influence future competition in telecommunications. Almost all states continue to regulate prices for local telephone services provided by the BOCs.[10] However, a few public utility commissions (PUCs) have decided that rate of return (ROR) regulation, in which profits for a BOC are determined by an allowed return on a historic measure of the asset base, may not be well suited for the increasingly competitive environment of telecommunications. The underlying rationale of ROR regulation creates (a) adverse incentives for a BOC to shift both costs and profits and (b) an attenuation of incentives for cost savings and innovation, because BOC profits will be largely unaffected by efficient economic behavior. Furthermore, the use of historical equipment costs as a basis for pricing may be extremely misleading in an industry in which new and improved technologies may dramatically alter the value of old technologies. With competition and ROR-set prices, the BOCs will be placed at a competitive disadvantage since their regulated prices will be considerably higher than the prices that competitors, using the new technology, can charge.

At least three alternatives to ROR regulation either have been adopted or are under active consideration. The first alternative, the social compact, fixes rates for, say, a five-year period for "essential" services and allows the BOC to establish unregulated prices for nonessential services. This alternative has been adopted by the Vermont legislature and PUC. The second alternative is the rate moratorium, adopted by New York, which freezes all rates upward over a given period but allows unregulated downward rate movement. Either approach gives the BOC a proportion of profits created by cost-saving efficiencies. The last alternative, the rate cap proposal, has been adopted in the United Kingdom and is under study at the FCC. It sets a maximum allowed change in prices, which depends on inflation and the rate of productivity change. However, individual prices are unregulated downward while being capped upward by the change in the average price index. Note that all these proposals create much greater price flexibility for regulated firms to meet emerging unregulated competition. Rather than the rates being set by regulators in a process in which competitors can create substantial obstacles, the telephone utilities will be able to benefit from gains in productivity and innovation and to set prices depending on competitive market conditions. However, under any of these changes the interaction of state regulation and federal antitrust laws will still create an uncertain situation, which is discussed by Sullivan in Chapter 4.

At the federal level the FCC is currently studying the adoption of a rate cap proposal for AT&T interLATA long-distance prices in place of ROR regulation. While AT&T long-distance services are subject to increasing competition, the FCC has continued to regulate AT&T under the "dominant firm" criterion. MCI and US Sprint, along with other long-distance competitors, have been effectively unregulated. However, AT&T has been regulated under the economic theory that AT&T retained market power in long-distance markets. The market power notion is that were regulation not present, AT&T would be able to raise its prices above competitive levels, at least for certain services and for certain customer groups. Thus, the competitive experiment has been constrained by FCC regulation of AT&T in long-distance markets. We think that decreased FCC regulation of AT&T is quite likely to occur in the near future. Future competition in long-distance markets will take place in a different competitive framework, one in which the constraints on AT&T's ability to set its own prices will change markedly.

Future Competition in Long-Distance Markets

The divestiture of AT&T basically arose from conditions in long-distance markets that emerged in the 1970s.[11] Changes in technology, both in microwave transmission and computerized central office switches, led to conditions in which competitors to AT&T's long-distance services emerged. After fifteen years of public antitrust challenges, regulatory disputes, and numerous private antitrust suits, the basic competitive structure of long-distance competition has been established. Three major networks, AT&T, MCI, and US Sprint, are largely in place, with a possible fourth network still to be established. The economic costs of the divestiture have been substantial—in the billions of dollars—to create equal access for the competing carriers as well as to administer the divestiture. Thus, the obvious question is whether this extremely expensive economic and legal experiment has been worth the cost.

To date, the experiment cannot be counted as an established success since AT&T continues to hold a very high market share. Furthermore, the earnings (and stock prices) of MCI and US Sprint have performed poorly over the past few years. Lastly, the largest proportion of the price decreases that have taken place in long-distance service since the MFJ have been directly caused by the lower costs of exchange access for AT&T that were created by the phasing out of the historic subsidy from long-distance service to local service. These decreases in AT&T's prices, decreed by the FCC, have "dragged along" the prices of MCI and US Sprint, which have been hard-pressed to maintain a 1 to 5 percent discount compared with AT&T prices. The situation thus differs markedly from the past, when MCI and US Sprint maintained substantial discounts under AT&T prices, aided by their 55 percent discount in exchange access costs before equal access was implemented. Indeed, MCI has publicly supported proposals to partly deregulate AT&T, a development that would be expected to lead to higher prices for some long-distance services. Thus, many observers agree with Huber's conclusion that "many, if not all" of AT&T's competitors remain in business at the sufferance of AT&T and the FCC and that AT&T still has "considerable market power."

In Chapter 5, "*United States v. AT&T:* An Interim Assessment," Roger G. Noll and Bruce M. Owen discuss possible reasons for the somewhat disappointing competitive performance to date. They emphasize the administrative implementation of equal access that made it difficult for competitors of AT&T to achieve significant gains in market share, and they discuss the difficulty of the competitors in establishing efficient national networks due to the slow pace of equal access conversions. While both factors have affected the competitive performance to date, the basic unanswered question is whether MCI and US Sprint can compete on a cost basis with AT&T in the future.

Before equal access, they could compete on cost because they received a substantial discount on their access costs. However, potential competition is typically difficult, if not impossible, to judge in a regulated context when competitors face different cost situations because of regulation. Furthermore, a post-MFJ technological development—extremely high capacity and low-cost fiber optic transmission capacity—may have changed the competitive framework against the challengers to AT&T because of its creation of large economies of scale. Huber discusses this development in Chapter 3, in which he states, "The future of competition in markets for long-distance services is under some shadow because of the revolution in fiber optics."

What about future competition in long-distance markets? Here the essential economic fact is that the fiber optic networks are in the ground. The fiber optic networks have substantial unused capacity, which will permit future competition without the necessity of substantial investment in additional facilities.[12] The investment costs of these networks are sunk and they will be used so long as their operating costs can be covered, even if investors in these facilities never achieve the rates of return they expected when they made their investments. Based mainly on this consideration, Noll and Owen are confident that competition "eventually will have worked by lowering prices and improving the efficiency of the long-distance market" even though the transition to competition may be quite lengthy.[13] Noll and Owen believe that MCI will remain a viable competitor because its cash flow will be sufficient to service its debt in the future.

The Noll and Owen analysis of the future survival of some competition to AT&T is unexceptionable.[14] They place their greatest emphasis on future competition arising from the large amounts of unused capacity that AT&T's competitors currently have. However, they conclude that even after the transition to competition has taken place, ". . . the best that can be expected in this industry is a dominant firm oligopoly with a handful of significant players. Moreover, we expect the advantages of AT&T to persist for years." But they expect this situation to be a large improvement over AT&T's position in the predivestiture marketplace.

We would find a dominant firm outcome with AT&T having a 50 to 75 percent market share and maintaining price-setting leadership in the future to be a disappointing competitive outcome. AT&T would be largely setting the price with some competitive pressure from MCI, US Sprint, or their successors. However, we are more confident that competition for large business customers in long-distance markets will be extremely competitive. A high degree of competition already exists, and improvements in transmission and switching in the future should lead to increased competition. Furthermore, many large businesses have already established partial or complete alternative telecommunications networks. In Chapter 7, "The Role of the Large Corporation in the Communications Market," James L. McKenney and H. Edward Nyce demonstrate the competitive importance of telecommunications by presenting a case study that describes the corporate goals of large businesses in telecommunications and the experience of Manufacturers Hanover Trust in establishing its own global telecommunications network. Also, here the available unused network capacity becomes an extremely important competitive factor because large corporations can coordinate their own network planning and purchasing to take advantage of the excess capacity.

However, we are much less confident regarding future competitive prospects for long-distance service to small businesses and residential customers. At present, marketing and administrative costs are quite important; they compose between 66 and 75 percent of total costs (excluding access charges). Thus, the unused network capacity is much less of a future competitive factor. MCI and US Sprint have so far been unsuccessful in achieving competitive inroads against AT&T in these markets. Indeed, MCI has reduced its competitive emphasis on these markets and increased its emphasis on large user markets. AT&T has current market power for residential and small business customers and is likely to continue to do so in the future.[15]

This situation has led some analysts to state that only BOC participation will provide competition to AT&T in the future, and it is only a matter of time before the BOCs return to long-distance markets. Huber calls this outcome the "most ominous competitive threat" to AT&T.[16] We would rather not speculate on this possibility, since it would be a sharp reversal of U.S. government policy. Furthermore, it would seem to require the replacement of Judge Greene's authority through the MFJ. Nevertheless, it has been argued that the establishment of equal access provides the necessary conditions to allow for BOC participation so that Judge Greene should relax the MFJ. We feel that Judge Greene is extremely unlikely to so decide at any time in the foreseeable future, since his condition appears to be the presence of substantial competition at the local level.

Thus, our evaluation is that current and future competition is likely to be quite high for large business customer long-distance services. Indeed, we see no reason for continued regulation of AT&T for these services. However, residential and small business customers currently present a markedly different competitive situation. We are not confident that competitive prospects for the future are significantly better, especially since express capacity will have less competitive importance, given the current competitive framework. Future actions by the FCC, by the Justice Department, and perhaps by Judge Greene may all have an important influence on future competition in residential and small business long-distance markets.

Future Competition in Local Services

Although the MFJ prohibited BOC participation in long-distance markets, the BOCs inherited the provision of local services from predivestiture AT&T. The "bottleneck monopoly" held by the BOCs was considered to arise from a "natural monopoly" that they held. That is, no one could provide local services at nearly so low a cost as the BOCs, so that competition for local services would not exist until future technological developments permitted competition at the level of the local loop, or "last mile," which connects customers to their local central office. Technological developments, fiber optic transmission, and vastly improved customer switches (PBXs), have occurred. They are currently affecting competition for local services and will have a greater influence in the future. However, for residential and small business customers this technological innovation has not noticeably affected competition to date.[17] Thus, the bottleneck that has led the U.S. government and Judge Greene to maintain the prohibition of BOC competitive entry into long-distance markets still remains.

Competitive conditions for large businesses at the local level are markedly different already, with even more competition likely to appear in the near future. While only minimal competition exists for local exchange calls, important competition already exists for exchange access service (connection to long-distance carriers) and competition for intraLATA long-distance has begun to emerge in many states where PUCs have permitted it. The reasons for this competition are technological. The decreased price and improved performance of digital PBXs have permitted large corporations to concentrate their traffic and to use high-capacity transmission options such as fiber optics, which permit a significant cost savings over the older, copper-based transmission systems.

Here we report on competition in Manhattan, which is the most advanced in the United States given the commencement of operation, in 1985, of Teleport, a fiber optic metropolitan area network (MAN) transmission system and competitor of New York Telephone.[18] However, a fiber optic MAN system, ICC, commenced operation in the District of Columbia early in 1987 and Chicago Fiber Optics began operation in the autumn of 1987. We expect other fiber optic systems to be constructed in the near future.

For large users, numerous demand substitutes exist for access to a long-distance carrier's point of presence (POP). Competition arises from alternative access carriers such as Teleport, from private microwave facilities, and from direct connection by interexchange carriers. However, the major competitive alternative to New York Telephone's special access facilities, used by large businesses in Manhattan, is Teleport. The amount of capacity offered by Teleport is an important competitive factor.

An estimate of Teleport's current competitive importance can be made by comparing Teleport's current supply of high-capacity DS-1 circuits, which connect to customers, with New York Telephone's supply. Using the Teleport estimate of 405 DS-1 circuits, Teleport currently supplies about 25 percent as many high-capacity circuits as New York Telephone does. The respective capacities supplied by Teleport and New York Telephone demonstrate that after only two years of operation, Teleport already has a substantial competitive presence in the provision of high-capacity circuits to large users. Furthermore, Teleport has sufficient capacity in place to increase its service by a factor of 35.

To determine the potential competitive importance of Teleport, we matched the current locations of the Teleport fiber optic network in Manhattan with New York Telephone's 400 largest customer locations. These customers are all sufficiently large to make use of Teleport DS-1 (T-1) facilities.

Furthermore, while those 400 customer locations represent

less than 1 percent of all New York Telephone business customer locations in Manhattan, they represent approximately 32 percent of the interexchange, interLATA business toll usage in Manhattan. The results demonstrate that 64.5 percent of the top 400 customer locations are in buildings adjacent to current Teleport routes. Alternatively, locations passed by Teleport routes account for 66.4 percent of interLATA toll usage for the top 400 customers. Thus, nearly two-thirds of the 400 largest customer locations in Manhattan currently have the competitive option of choosing Teleport Communications over New York Telephone for access to interexchange carriers.[19]

To estimate the amount of current use of alternative facilities to New York Telephone, a survey of large users was conducted in March 1987. The 500 largest New York Telephone customers in Manhattan were surveyed about their telecommunications network uses. Of the respondents, 32 percent report private long-distance networks, non–New York Telephone access services, or other forms of competitive alternatives to New York Telephone services. Thus, a significant number of large users in the New York City metropolitan LATA currently use services competitive to those offered by New York Telephone.

The use of the alternative access to interexchange carrier POPs has already had a significant competitive effect (for large business customers) on New York Telephone's revenues. An econometric analysis of the survey data, together with New York Telephone usage data, allows us to determine the effect of alternative exchange access for the large users.[20] For those companies which had alternative exchange access to New York Telephone, the growth rate of use of New York Telephone exchange access was about 35 percent slower than for comparison companies without alternative exchange access. The results are highly significant statistically. Thus, among the 500 largest New York Telephone customers, the 35 percent decrease in rates of growth yields an estimate of 8.5 percent slower rate of growth in exchange access among New York Telephone's 500 largest customers in Manhattan due to competitive alternatives for exchange access.

Currently, New York Telephone's rate for a two-mile interoffice DS-1 line, $1,062 per month, is quite close to the price charged by Teleport, $1,006.[21] Similarly, the prices charged by Manhattan Cable Television and LOCATE, two additional competitors to New York Telephone, are between \$800 and \$1,000 per month. However, New York Telephone has recently applied to the FCC for permission to decrease its prices by about 20 percent, and permission is likely to be granted. Thus, increased competition is the likely outcome, with falling prices for high-capacity transmission services for large business customers in New York City. Together with the high degree of competition for large business customers among the long-distance companies discussed in the previous section, we expect overall telecommunications competition to increase at both the local and the long-distance levels for large customers. This increase in competition will occur in all large urban areas, not just New York City, as the improved technology becomes more widespread.[22] Thus, the current legal framework established by the MFJ may be hindering rather than encouraging competition for telecommunications services provided to large business customers.

Grandon Gill, F. Warren McFarlan, and James P. O'Neill, in Chapter 8, "Bypass of Local Operating Telephone Companies: Opportunities and Policy Issues," support the analysis of increased competition from bypass companies in all major metropolitan areas. They argue that there is a first-mover advantage to be gained by these companies and that the financial community is likely to consider them attractive investments. Later these authors would expect a period of consolidation through mergers and acquisitions, although competition with the BOCs would remain strong.

Future Long-Distance Competition for Residential and Small Business Customers

Although competition is well established for large business customers, competition on the local level is minimal for residential and small business customers and much less well developed among long-distance services. The main technological and economic reason for this lack of competition is that residential and small business customers do not generate sufficient traffic to make use of high-capacity transmission circuits or computer switching. Indeed, these customers depend on the BOC-provided central office switch as their outlet to the outside world. We believe that for these customers increases in local competition are unlikely to occur in the near future. But increased long-distance competition could soon be created by a change in government policy.

Large businesses are able to achieve very competitive long-distance service by utilizing their PBXs to provide least-cost routing. For any given call to a given city at a given time of day, a computer program chooses the long-distance service that costs the least, taking into account the myriad of quantity discounts, time of day discounts, and other special plans offered by the long-distance competitors. Small businesses and residential customers will have only a single long-distance carrier, so that they cannot engage in least-cost selection of long-distance calls.

However, small business and residential customers could be provided with least-cost routing through their *public* exchange, the BOC central office switch, rather than through the *private* branch exchange (PBX), which large business customers use. Judge Greene has interpreted the MFJ to prohibit BOC provision of least-cost routing under the MFJ. A change in government policy would likely increase competition dramatically; as one AT&T official stated, it would turn the long-distance business into a "commodity business" overnight. The problems that AT&T's competitors have not overcome in successfully marketing their services to residential and small business customers would be of much less importance, and the current unused capacity of their networks would be likely to lead to increased price competition, in an outcome similar to what has occurred for large business customers.

Thus, to date competition in telecommunications services, both long-distance and local, has increased for large business customers since the divestiture of AT&T. Certainly for long-distance service and also increasingly for local high-capacity service, competitive alternatives exist and are widely used. Regulatory and legal measures that would lead to increased competition for small business and residential customers seem called for.

Future Competition in Telecommunications Equipment

Almost all telecommunications equipment markets are very competitive and should remain so in the future. Until the early 1970s, the BOCs bought almost all their equipment from AT&T's manufacturing arm, Western Electric. They rented the equipment to end users or used the equipment internally. Since the BOCs supplied about 80 percent of the telephone lines in the United States, the large majority of U.S. telephone equipment markets were not open to competition.[23] Yet today the BOCs purchase their own equipment and sell equipment from a number of U.S. and international suppliers. What has led to this markedly more competitive situation? Our view is that changes in neither regulation nor the MFJ deserve credit here; rather, changes in technology are the most important reason for increased competition.

Huber discusses this view in Chapter 3, "The Technological Imperative for Competition."[24] The development of semiconductor and integrated circuit technology has moved considerable amounts of network intelligence from the BOC central office to the business end user in the form of digital PBXs and intelligent terminals. By the early 1970s, the FCC permitted competition in these markets, and the "interconnect" industry rapidly came into being. By the early 1980s, at the time of the MFJ, AT&T had less than a 25 percent share of PBXs, for instance. Currently the MFJ allows the BOCs to sell customer premise equipment (CPE) but not to manufacture it because of the II(D)2 restriction of the MFJ. Numerous companies, such as AT&T, Northern Telecom, NEC, and IBM, both manufacture and sell CPE. Competition is extremely high, as the recent shakeout that the CPE markets are experiencing demonstrates. CPE has become an international market, with telecommunications and computer manufacturers competing on both price and technology dimensions.

Central office switches are the main piece of equipment used internally by the BOCs. Because Western Electric and Bell Laboratories decided not to produce digital central office switches, AT&T had already lost 50 percent or more of the central office switch market, mostly to Northern Telecom, the manufacturing arm of Bell Canada, at the time of divestiture. AT&T has made a comeback with its own digital central office switch and now sells about the same amount as Northern Telecom does. Although European companies and NEC are attempting to sell switches to the BOCs, so far they have met with limited success. In Chapter 6, "The Future Evolution of the Central Office Switch Industry," Jerry A. Hausman and Elon Kohlberg discuss the likely future evolution of the industry. Here competition is global in scope, investments in R&D are extremely high, and the number of competitors is decreasing. Nevertheless, Hausman and Kohlberg foresee an increasing level of competition as foreign competitors use government subsidies to compete with AT&T and Northern Telecom.

However, these authors also raise possible long-term problems that may arise from the structure of the U.S. telecommunications industry as put into place by the MFJ. That telecommunication markets are closely related may affect a company's competitive position and long-term strategy. AT&T both provides long-distance services and manufactures central office switches. Hausman and Kohlberg predict increasing competition in the provision of telecommunications services among the BOCs and AT&T. They do not believe that the BOCs will want to remain dependent on an unregulated competitor for the technology that will be required to provide advanced telecommunications services. Thus, they conclude that the structure of the MFJ may lead the BOCs to turn increasingly to foreign suppliers of central office switches in the future unless some adjustment in AT&T's corporate structure is made.

Lastly, we consider the possible future impact of Judge Greene's September 1987 decision to continue to exclude the BOCs from manufacturing telecommunications equipment despite Justice Department and FCC recommendations to the contrary. Interpreting Judge Greene's decision as nonlawyers, we find somewhat puzzling his application of the antitrust law to continue BOC exclusion from manufacturing. The BOC purchases are small in comparison to the size of the telecommunication equipment markets, which are either national or international in scope. Nor is it likely that potential BOC anticompetitive behavior could distort competition in these markets, as has been demonstrated by successful BOC participation in the CPE markets since divestiture. However, even without BOC participation telecommunications markets will continue to be quite competitive. The economic losses that will arise from BOC exclusion from these markets are probably not overwhelming, given the amount of competition that currently exists.

Future Competition in Information Services

Information services arise from the interaction of computers with telecommunications. Videotext services, audiotext services, and voice-messaging services all provide examples of information services in which the MFJ restrictions, along with FCC regulation, have led to significant losses to the U.S. economy. For instance, videotext services have to date been a business failure in the United States. Trintex, a joint venture of IBM and Sears, Roebuck, is the latest videotext enterprise scheduled to begin operation. Numerous other videotext efforts have failed, and CBS has already exited from Trintex. Yet France has an extremely successful videotext system with more than 2 million subscribers and a growth rate exceeding 1 million subscribers per year. This system, called Minitel, is operated by the (centralized) French phone system and private suppliers who provide the videotext content.[25]

The FCC has changed its regulatory policies to permit BOC participation in information service markets so long as the BOCs provide a type of "equal access" to competitors in the field. The Department of Justice, the NTIA in the Department of Commerce, and the FCC all recommended to Judge Greene that he remove the MFJ restrictions on BOC participation in information service markets in the triennial review of the MFJ. Judge Greene turned down this recommendation, although he did significantly relax the MFJ restrictions. In his September 1987 and March 1988 opinions he permitted BOC provision of information service network infrastructures and gateways. However, all "information services [will be] originated by others," so that the BOCs will be excluded from providing information service content. They will not, for instance, be allowed to compete in the provision of electronic Yellow Pages.

Here we believe that future competition may be greatly affected by Judge Greene's decisions. Large potential losses to the U.S. economy may occur if information services are not well developed. The generation, transmission, and use of computerized information are an extremely important competitive factor for a host of U.S. businesses, both large and small. Given the network requirements for information services and other associated scale economies, along with the use of the central office switch to provide advanced information services, BOC participation in these markets seems crucial. Coherent government policies that will lead to increased provision of information services and competition among information service providers are needed. To date, the effect of the MFJ, which is rooted in the technological situation of the 1970s rather than the 1980s or 1990s, has been extremely adverse for the provision of and competition in information services.

The problems that the BOCs have faced to date and need to solve for future competition are discussed by Richard L. Nolan, C. Rudy Puryear, and Dan H. Elron in Chapter 10, "The Hidden Barrier to the Bell Operating Companies and Their Regional Holdings Companies' Competitive Strategies." These authors contend that the BOCs have not yet made the transformation to the "high-tech" status that the semiconductor and computer developments have created. If so, the ability of the BOCs to compete in advanced information service markets may be limited until they transform their organizations. In Chapter 9, "Enhanced Communications Services: An Analysis of AT&T's Competitive Position," Joseph Baylock, Stephen P. Bradley, and Eric K. Clemons consider AT&T as a possible future competitor in information service markets. Although the MFJ restricts AT&T from full participation in information service markets until at least 1989, the company's previous efforts in providing networks for computer applications have been quite unsuccessful. AT&T is an important potential competitor, especially given the advent of fiber optic networks, in which transmission costs are an extremely small fraction of overall information service costs.[26] However, Baylock, Bradley, and Clemons acknowledge that although AT&T has great expertise in the key enabling technologies required for information services, those technologies will be widely available from a competitive marketplace and AT&T lacks the specific industry knowledge and certain business skills to be a leader in information services, other than as a delivery vehicle.

Conclusion

Competition for long-distance service provision to large businesses exists now and will continue to exist in the future. Much less competition exists in long-distance markets for small business and residential customers; indeed, AT&T seems to have significant amounts of market power for these customers. Thus, the keystone of the divestiture of AT&T through the MFJ can be judged only a partial success.

Competition in telecommunications equipment markets is

quite high, yet significant competition already existed prior to the MFJ and there is no reason to expect any decrease in competition. The interaction of telecommunications equipment and computer equipment will continue to evolve, with advanced functionality and services creating increasing benefits to consumers and to the U.S. economy.

Provision of information services and competition in information service markets have been substantially hindered by the MFJ. Information services are crucial to the future competitiveness of the U.S. economy. In his September 1987 and March 1988 decisions, Judge Greene went part way in allowing limited BOC participation in information service markets. The future of economical and more diverse information services is likely to be the most important development in U.S. telecommunications markets over the next decade. If such a development occurs, wide-ranging effects on the U.S. economy, from manufacturing to financial services to home entertainment, can be expected to occur.

Notes

1. See the amicus curiae brief filed by the FCC on April 20, 1982.
2. This position was supported by the BOCs and many, although not all, state regulators.
3. Jerry A. Hausman, a coauthor of this chapter, filed an affidavit to the district court on behalf of Pacific Telesis, one of the seven regional holding companies.
4. Of particular interest is that both sides of *United States v. AT&T* have now apparently switched sides of the debate. AT&T, which vigorously fought against divestiture, now takes the position advocated by the Justice Department in the early 1980s, at least with respect to manufacturing and long-distance services.
5. Numerous congressional bills that would revise the Communications Act of 1934 have been introduced and debated over the past decade. The latest unsuccessful attempt was the Dole bill, introduced in 1986.
6. This statement holds for local exchange and exchange access services. However, widely divergent expectations existed regarding the possibility of competition for intraLATA long-distance services.
7. The Justice Department takes the position that this incentive problem is a regulatory one, best handled by regulators, unless the antitrust laws are violated by predatory pricing by the BOCs. Predatory pricing by the BOCs is extremely unlikely, given competitors such as AT&T in the interexchange markets. However, Judge Greene disagrees with the Justice Department's position. Despite numerous attempts by both the Justice Department and private plaintiffs to prove that AT&T had violated the antitrust laws through predatory pricing, AT&T was never found guilty of the charge.
8. Under the "essential facilities" doctrine, AT&T was found to have abused

its local bottleneck monopolies during the 1970s in its competition with MCI. *MCI Communications v. AT&T* (7th Cir. 1983), 708 F.2d 1081. Kevin Sullivan in Chapter 4 discusses the legal background of this doctrine.

9. *United States v. Western Electric Cl., Inc., et al.; Notice of Entry of Final Judgment,* 47 Fed. Reg. 40392 (Sept. 13, 1982).

10. However, a number of state PUCs have deregulated intrastate, interLATA long-distance service that they have jurisdiction over. After divestiture, AT&T was typically regulated for these services, although MCI and US Sprint were often unregulated.

11. While both the 1949 antitrust suit brought against AT&T, which was settled in 1956, and the 1974 antitrust suit, which led to the MFJ and divestiture, had their primary origins in telecommunications equipment markets, the emergence of MCI and other long-distance competitors, along with AT&T's reactions to this competition, was the main cause of the divestiture.

12. Electronics must often be added to make the fiber optic capacity usable. However, the price of the necessary electronics has been decreasing extremely rapidly over the past five years.

13. Note that the implicit price and efficiency comparison should be to the counterfactual situation of future FCC regulation of monopoly AT&T provision of long-distance services. At the present time FCC regulation of AT&T prices is the basic price-setting mechanism in most long-distance markets. Under a highly competitive situation when marginal costs are far less than prices and substantial unused capacity exists, AT&T's competitors might be expected to decrease prices in an attempt to gain increased demand and market share.

14. However, we would not be surprised to see MCI and US Sprint merge their operations or be taken over in the future.

15. The effects of the greater degree of competition for business customers in comparison with residential customers is evident in AT&T's November 1987 proposal to the FCC for changes in switched long-distance rates. AT&T was quoted to the effect that "day rate period margins are too high, and its discount rate period margins are too low, to continue to be sustained in a competitive marketplace. The seriousness of this problem has increased, moreover, as competition in the interexchange industry has intensified in recent years" (*Telecommunications Reports,* November 23, 1987, p. 8). The great majority of business class demand is during the day rate period, while residential customers are the largest users of long-distance service during discount periods. AT&T has proposed a decrease in day usage rates of 6.3 percent, together with a reduction in the evening period discount off the day rate from 38 to 35 percent and a reduction in the night/weekend discount from 53 to 50 percent.

16. See Peter Huber, "The Technological Imperative for Competition," Chapter 3 of this volume.

17. Roger G. Noll and Bruce M. Owen raise the possibility that cellular radio or some other radio technology may allow for competition to the local loop. We have studied this matter previously and believe that this outcome is extremely unlikely. The reverse is more likely to happen—future broad-band information services will require even more transmission capacity than the twisted wire pairs of the local loop currently provide. See also Chapter 3, in which Huber concludes that there is "no likelihood that airwave transmission systems will soon offer an economically viable access substitute for twisted copper wire."

18. For a more complete discussion, see J.A. Hausman, T.J. Tardiff, and H. Ware, "Competitive Telecommunications Markets for Large Users in New

York," mimeo, December 1987; and Huber, "The Technological Imperative for Competition," Chapter 3 of this volume.

19. Manhattan Cable Television provides coaxial cable-based transmission competition to New York Telephone, while LOCATE provides microwave-based transmission competition. Both services have prices that are quite close to New York Telephone prices. Large business users can also install their own fiber optic, coaxial, or microwave-based transmission systems. A survey by the New York Clearinghouse Association, whose members are large New York City banks, indicates that the majority have installed private systems. See also Chapter 7, in which James L. McKenney and H. Edward Nyce describe the development of Manufacturers Hanover's private network.

20. For further details see Hausman, Tardiff, and Ware, "Competitive Tele-communications Markets."

21. This price is the average of the quoted Teleport prices for one-zone and two-zone crossings.

22. In January 1988, Teleport announced a planned expansion to twenty-one additional cities. Construction by Teleport of an alternative system for large users in Boston is currently under way.

23. The situation in which the BOCs purchased almost all their equipment from Western Electric was the focus of the 1949 Department of Justice anti-trust suit against AT&T and provided the origin of the 1974 Justice Department action that led to the MFJ.

24. Noll and Owen agree with this view in their discussion of telephone equipment markets; see Chapter 5.

25. Germany also has a successful videotext system, while systems in Japan and the United Kingdom have been less successful. Some analysts argue that the French system is unlikely to be commercially successful but exists mainly because of government subsidy.

26. Indeed, AT&T announced new information services for large users in April 1988. These large users are served over fiber optic networks with low local-access transmission costs.

Part One
Industry Perspectives

2
AT&T and the Public Good: Regulation and Competition in Telecommunications, 1910–1987

Richard H.K. Vietor

Effective, aggressive competition, and regulation and control are inconsistent with each other, and cannot be had at the same time.
—Theodore Vail, 1910

In the age of Theodore Vail, AT&T adopted a strategy of providing universal service through an end-to-end, national monopoly. Eventually the government developed pricing and cost allocation systems that harnessed the Bell System's technological innovation to serve these very goals. This approach worked phenomenally well until the late 1960s. By then, however, the distortions of regulatory cross-subsidy diverged too far from the economics of technological change. Under such pressure, the social consensus in telecommunications policy crumbled.

This chapter presents a historical overview of regulatory change and competition in telephonic communications. Chronologically, the story has three parts. In the first part, from the turn of the century to 1956, a system of economic regulation by government evolved at the state and federal levels. The goal of this regulation, which Congress more or less made explicit in the Communications Act of 1934, was "universal service." Thereafter, by controlling entry, price, facilities, and product

Note: The author is grateful for the helpful criticisms of Davis Dyer, Alan Gardner, Alfred Kahn, Robert Lewis, Louise McCarren, John McGarrity, F.F. Stoddard, and Peter Temin.

offerings, regulation shaped industry structure and defined the boundaries of telecommunications markets.

In the second part of the story, from 1956 to 1984, the fabric of this regulatory system was gradually unraveled, almost imperceptibly at first, by increasing competition. Over AT&T's persistent opposition the Federal Communications Commission, under pressure from federal judges, allowed limited entry into the markets for customer premise equipment and long-distance (interexchange) service. There were several reasons for this regulatory reform: rapid technological innovation, ambitious entrepreneurs, new economic conditions and political norms, new ideas, and regulatory failure. The process accelerated in the mid-1970s until it seemed to spin out of institutional control. This transition, from regulated monopoly based on electromechanical technology to regulated competition based on electronic digital technology, culminated in the breakup of the Bell System on January 1, 1984.[1]

Three chief executives, each with a distinctive approach, struggled to manage the transition for AT&T. Under the leadership of H. I. Rommes after 1966, the Bell System tried to accommodate a degree of competition at the outskirts of its monopoly network. When John deButts took over in 1972, the company called for a moratorium on experiments with competition, urging the polity to assess the implications of change. In 1979, after this approach had run its course without success, Charles Brown took over. Haltingly at first, Brown acceded first to competition and finally to divestiture in exchange for a vision of the information age.[2]

The third part of the story, a saga that continues today, begins with divestiture and the dilemma of regulated competition. Breaking up the Bell System may have created more problems than it solved; taken together, equal access (to the local exchange) for long-distance companies and providers of enhanced (computer-related) services, asymmetrical regulation, efficient cost allocation and pricing, and the Modified Final Judgment itself pose an extraordinarily complex public policy puzzle.

No story that I know of provides a richer, more dynamic opportunity for studying the linkages among technology, economics, politics, and corporate strategy than the history of policy in telecommunications.

Regulation-Defined Markets

The Origins of Universal Service

Writing in the *Annual Report of 1910,* Theodore Vail, chairman of AT&T, summarized his company's strategic objective as follows:

> The position of the Bell System is well known. . . . The telephone system should be universal, interdependent and intercommunicating, affording opportunity for any subscriber of any exchange to communicate with any other subscriber of any other exchange . . . *annihilating time or distance by use of electrical transmission.*[3]

To attain this goal, Vail had recently taken the radical position of publicly endorsing regulation by state government. Regulation, Vail had concluded, was the only politically acceptable way for AT&T to monopolize telephony and achieve the network externalities and economies of scale inherent in the technology.

Since 1876, when Alexander Bell invented the telephone, the market for telecommunications had grown immensely. Technological innovation continued at a rapid pace, making possible switched networks (organized in local exchanges), improved quality of transmission, and greater distances.[4] This progress seemed relatively consistent, despite alternating bouts of competition and monopoly.[5] (See Table 2-1.)

The Bell interests were originally organized as a patent association. Manufacturers were licensed, phones and private lines were leased, and local exchange companies were franchised throughout the country. As local franchisees grew capital constrained under the constant pressures for rapid expansion, the Bell interests extended their ownership, hesitantly at first, into the local exchange business. Increasingly, pressures for technical standardization and network coordination propelled the company toward a strategy of horizontal integration.[6]

During the first few years, the Western Union Company vigorously contested both the field and the Bell company's patents. But in 1879, the two companies agreed to settle their rivalry. Western Union withdrew from the telephone segment of the market, presumably to protect its position in long-distance business communications. Although telephone technology appeared unlikely to challenge the intercity capabilities of telegraph, this surrender now seems incredibly myopic.

Table 2-1
Growth of Telephone Market, 1876–1910

	Number of Telephones (000)	CAGR	Number of Calls per Day (000)	CAGR	Number of Companies	Bell System Share (%)[b]
1876	2.6		237		1	100
		31		13		
1893	266		1,906		18	100
		22		19		
1910	7,635		36,161		3,100[a]	52

CAGR = compound annual growth rate.
[a]3,100 (1902, in *Business Census*); for 1913, MacMeal reports 20,000 independent exchanges.
[b]Pertains to both number of telephones and number of calls.
Sources: *Historical Statistics of the United States*, Series R 1–12; and H.B. MacMeal, *The Story of Independent Telephony* (IPTA, 1934), pp. 204, 267.

For the next fourteen years the Bell System had a virtual monopoly, secured by its increasing control of communications' patents. Yet the sourcing of telephone handsets and switching equipment from a variety of manufacturers remained a major bottleneck. There were problems with capacity, quality control, technical uniformity, coordination, and competition for franchisees among Bell suppliers. In 1882 the Bell interests solved these problems by acquiring Western Electric, the largest manufacturer of electrical equipment in America. If vertical integration was not explicitly the Bell strategy, it soon would be.[7]

In 1885, Theodore Vail resigned as general manager of American Bell to organize a wholly owned long-distance subsidiary, incorporated as the American Telephone & Telegraph Company. Vail already had an extraordinary vision, to "make the powers of this Company to build . . . lines extending from any city in the state, to each and every other city . . . in the U.S., Canada and Mexico."[8]

After less than a decade, the fundamental elements of AT&T's business strategy were already in place: horizontal integration of local exchanges, backward integration (into equipment manufacturing), forward integration (through the practice of retail equipment leasing), and first-mover development of an interexchange (long-distance) transmission network.

Despite these preparations by the Bell System and despite its control of many key patents, entry and competition spread rapaciously after 1894, when the two fundamental patents expired. Significant opportunities for profit under the Bell pricing umbrella of leased telephones and private lines lured more than 3,000 commercial entrants during the next nine years. Competing (and nonconnecting) telephone service developed in more than half of all cities with populations larger than 5,000.[9] Hundreds of mutual systems and "farm lines" were also organized by rural entrepreneurs, either to avoid control by the Bell System or to introduce service where Bell was uninterested.[10] To supply these concerns with equipment, a host of manufacturers entered the market, including Automatic Manufacturing, Stromberg-Carlson, and Kellogg Switchboard and Supply.[11] Only long-distance service, for which the Bell System controlled critical patents, remained impervious to sustainable competitive entry, despite several cooperative efforts by independents.[12]

The Bell System employed a host of aggressive practices to oppose entry and competition. Among these practices were (a) a public relations campaign against independents, (b) refusal to connect with many independents (and all competitive systems), (c) refusal to sell telephone instruments to non-Bell interests, and (d) various efforts to break up or financially control those interests which threatened to establish significant networks or cooperative toll systems.[13] Refusal to interconnect was key. Without interconnect, an independent system could connect with neither Bell System exchanges nor other nonadjacent independent exchanges for which AT&T Long Lines provided toll service. With opportunities for growth limited, independent systems could not readily compete with the local Bell operating companies.

In 1907, Theodore Vail returned to become chairman of AT&T (after two decades away from the company as a venture capitalist). Vail curtailed the policies of direct competition, such as fighting entry with new construction and not selling equipment to independents. And he substantially liberalized the company's policy of sublicensing independent exchange operators, effectively dividing the independent movement. At the same time, he accelerated the rate of acquisitions, especially in the Midwest, with Morgan financial backing.[14] "Wherever it could be legally done, and done with the acquiescence of the public," wrote Vail in 1910, "opposition companies have been acquired and merged into the Bell System."[15]

This acceleration of takeovers, together with the backing of Morgan financial interests, precipitated pressure for federal regulation. Although state regulation was already widespread, no state commission had either the resources or the jurisdictional authority to control AT&T. Telephone and telegraph were deemed public utilities and, for the most part, monopolies. As one congressman typically complained, "To leave them unregulated is to leave private corporations with the power to tax the people at will."[16] In 1910, when Congress passed the Mann-Elkins Act to reform railroad regulation, it extended the Interstate Commerce Commission's authority to include telecommunications. Telephone and telegraph carriers were declared common carriers, whose rates were to be "just and reasonable."

The Mann-Elkins Act did not, however, address industry structure or competition in telecommunications. On the one hand, the public (and its regulators) chaffed at the network inefficiencies that resulted from competition. More than half of all Bell exchanges in cities with a population exceeding 5,000 faced direct competition. In state after state, public utility commissions reached the same conclusion:

> Competition resulted in duplication of investment, the necessity for the business man maintaining two or more telephones, economic waste to the company, increased burden, and consequent continuous loss to the subscriber. The policy of the state was to eliminate this by eliminating as far as possible, duplication.[17]

Between 1907 and 1913, twenty-six states enacted laws making physical interconnection compulsory.[18] On the other hand, antimonopoly sentiment was intensifying, especially with agitation from independents. For Vail, this contradiction was self-evident:

> Effective, aggressive competition, and regulation and control are inconsistent with each other, and cannot be had at the same time.
>
> Control or regulation, to be effective, means publicity; it means semi-public discussion and consideration before action; it means deliberation, non-discrimination.
>
> Competition—aggressive, effective competition—means *strife*, individual warfare; it means contention; it oftentimes means taking advantage of or resorting to any means that the conscience of the contestants or the degrees of the enforcement of the laws will permit.[19]

Vail chose at this time to put AT&T squarely behind government regulation, as the quid pro quo for avoiding competition.

In response to widespread complaints, the Justice Department began investigating AT&T's competitive practices in 1912. The company agreed to suspend pending acquisitions until the issues were resolved, hoping to obtain the Justice Department's informal approval on a case-by-case basis. But in March 1913, James McReynolds, attorney general for the incoming Wilson administration, gave notice that informal arrangements were unacceptable. Instead, McReynolds urged the Interstate Commerce Commission to assert its authority. In July, he filed an antitrust suit against Pacific Telephone for acquiring the Northwestern Long Distance Company.[20]

After several months of negotiations, the Justice Department and AT&T reached a historic agreement, commonly called the Kingsbury Commitment. In December 1913, Nathan Kingsbury, an AT&T vice president, outlined the terms of this agreement in a letter to the attorney general. There were three key provisions. First, AT&T would dispose of Western Union, which it had acquired in 1909. Second, AT&T would cease acquiring independent telephone companies "operated in competition with the Bell System." And third, the company would henceforth allow "qualified" interconnection with the Bell System for interexchange and long-distance service. Both Attorney General McReynolds and President Wilson expressed appreciation for the frankness and good faith of AT&T.[21]

The Kingsbury Commitment effectively redefined telecommunications markets and reshaped industry structure. In contrast to the breakup of Standard Oil just two years earlier, the vertically integrated Bell System was left intact and free to extend its geographic scope by acquiring noncompeting companies. During the next five years, the Bell System increased its share of the market from 58 to 63 percent (acquiring 240,000 stations from independents and giving up only 59,000).[22] Vail's strategy for pursuing universal service had received a tremendous boost.

In mid-1918, the government exercised its wartime prerogative to assume direct control over the telephone network. Ironically, the pressure for greater network efficiency during this period ran squarely up against the Justice Department's prohibition on consolidation of competing exchanges. A.S. Burleson, the postmaster general with wartime responsibility for communications, sought to eliminate competition and integrate operations wherever possible.[23] He also pushed disparate toll prices (on a mileage basis) closer to national uniformity.

After the war, pressure mounted from state regulators, the Bell companies, and even the independents to undo the Kingsbury Commitment's prohibition on consolidation of competing exchanges. Congress obliged by passing the Willis-Graham Act in 1921, explicitly permitting consolidation of competing telephone companies. Competition, evidently, was out of vogue. "It is believed to be better policy," urged Representative Graham, "to have one telephone system in a community that serves all the people, even though it may be at an advanced rate, properly regulated by State boards of commissions, than it is to have two competing telephone systems."[24] The Senate Commerce Committee simply asserted that "telephoning is a natural monopoly."[25]

The Bell System immediately launched an expansive program of acquisition—net purchases of 114,000 stations in 1921 alone. By 1930, the Bell System had acquired more than 250 independent companies, bringing in 1.4 million more stations (while divesting only 150,000). Its market share increased to 79 percent.

To prevent a political backlash, AT&T adopted a cooperative posture toward the independents. This strategy, which continued throughout the twenties, took several forms. First, AT&T agreed to notify the United States Independent Telephone Association in advance of each acquisition. Such notification allowed affected companies to intervene, possibly making the purchase themselves.[26] Second, the company provided extensive help to independents with which it interconnected: assistance with accounting and rate filings, legal and financial advice, provision of materials at cost, engineering advice, and even help with construction. Third, AT&T supported independents' efforts to obtain higher rates for local exchange services. While Bell operating companies obviously benefited from this support, it especially helped moderate claims on AT&T Long Lines for larger shares of joint long-distance revenues.[27]

During the 1920s, public utility commissions throughout the country adopted value-of-service pricing and statewide average rate making. Under the value-of-service concept (a precursor to Ramsey pricing), business users paid more than residential cus-

Table 2-2
Growth of Telephone Market, 1910–1930

	Number of Telephones (000)	CAGR	Number of Calls per Day (000)	CAGR	Number of Companies	Bell System Share (%)
1910	7,635		36,161		3,100	52
		5.0		4.9		
1930	20,103		83,520		6,000	79

Sources: *Historical Statistics of the United States,* Series R 1–12; and H.B. MacMeal, *The Story of Independent Telephony* (IPTA, 1934), pp. 204, 267.

tomers, since the benefit of service to business users was greater. Likewise, rates were higher in large exchanges (despite lower costs) than in small ones, since service (the number of possible connections) in large areas was superior. Similarly, statewide averaging of rates (for like-sized exchanges and toll calls of equal distance) appealed to public utility commissions on several counts: it encouraged new residential service through cross-subsidization, simplified administrative procedure, and gave the impression of fairness.[28]

During the Vail era, the market for telecommunications experienced steady growth (see Table 2-2), and for AT&T, solid profitability (Chart 2-1). Consolidation and interconnection increased calling opportunities, while periodic rate reductions further stimulated demand. But as a national network emerged, state regulators found it increasingly difficult to deal with the national monopoly that managed it. There was simply no effective way to supervise AT&T's operations, or separate its assets, along state jurisdictional lines.

National Monopoly, Federal Regulation

The Great Depression caused a major setback for the growth of telecommunications markets. Between 1929 and 1933, the number of telephones in use declined by 18 percent; residential subscribership dropped 25 percent, rolling back the household penetration rate to less than one-third. Long-distance usage, apparently the most income-elastic service, dropped 36 percent.[29] While this performance was certainly no worse than that in other sectors of the economy, it was significant politically, since the public increasingly viewed telephone service as a

Chart 2-1
Bell Telephone System Per Station Statistics from Company Records

1885–1894 EARLY PERIOD OF MONOPOLY
1894–1913 PERIOD OF COMPETITION
1913–1921 PERIOD OF THE KINGSBURY COMMITMENT
1921–1935 PERIOD UNDER WILLIS GRAHAM ACT

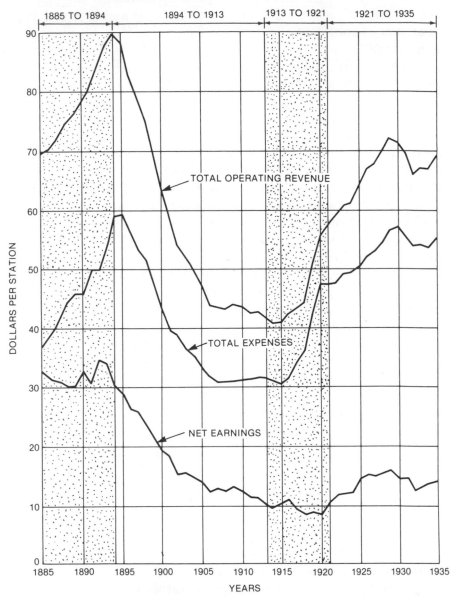

Source: FCC, *Investigation of the Telephone Industry in the United States,* in U.S. Congress, *House Document No. 340* (76th Cong., 1st Sess.), 1939, p. 135.

social necessity. AT&T's management further damaged its public image by cutting only costs, but not dividends, in the face of declining revenues. More than 150,000 employees were laid off.[30]

Although these problems no doubt contributed to the pressures for federal control, they were not unusual amidst the pervasive loss of confidence in private enterprise that permeated the national consciousness in the early 1930s. Big business everywhere came under attack and criticism for the massive failings of the economy. As confident young Braintrusters conceived of new ways to control securities markets and reorganize utility holding companies, telephony was scarcely overlooked.[31] For nowhere was business any bigger, or more closely linked to the Money Trust, than at AT&T—the world's largest corporation.

By the early thirties, AT&T was all but a monopoly. The Bell System served about 80 percent of local exchange customers (with 90 percent of local exchange volume); Western Electric did 92 percent of all telephone equipment sales; and AT&T Long Lines provided virtually 100 percent of all long-distance service.[32] (See Table 2-3.) This position of extraordinary dominance stood in stark contrast to the passivity of federal regulatory authority, as well as the lack of resources among state commissions. In the twenty-four years since the Mann-Elkins Act, the Interstate Commerce Commission never established an office to supervise telephone or telegraph services, never performed a valuation of telephone industry properties, and never conducted a general rate investigation.[33]

In December 1931, as part of a general investigation of American business, Congress passed a resolution calling for a study of the corporate structure and organization of the telecommunications industry. Dr. Walter M.W. Splawn was appointed to supervise the project, along with parallel studies of electric power and natural gas. In the spring of 1934, Splawn submitted his preliminary findings to the House Commerce Committee. Although he was as critical of the holding company structure in telecommunications as in other utilities, he did not suggest any major change in industry structure. Splawn concluded that, unlike the power utilities, the "telephone business is a monopoly—it is supposed to be regulated." Yet, "thus far, regulation, particularly by the federal government, ha[d] been nominal." Accord-

Table 2-3
The Bell System, Non-Bell Groups, and Independent
Telephone Companies at December 31, 1932[1]

Basis of Comparison	Total All Companies[2]	Bell System Companies		Companies Comprising the 38 Non-Bell Groups		The 69 Independent Telephone Companies[3]	
	Total	Total	Percent	Total	Percent	Total	Percent
Investment in plant and equipment	$4,660,662,997	$4,301,613,801	92.30	$322,735,806	6.92	$36,313,390	0.78
Operating revenues	1,049,757,095	989,722,645	94.28	52,350,694	4.99	7,683,756	0.73
Total miles of all wire	88,303,231	84,309,177	95.48	3,551,240	4.02	442,814	0.50
Average number of local exchange messages originated per month	2,323,872,236	2,087,766,344	89.84	193,788,480	8.34	42,317,412	1.82
Average number of toll messages originated per month	70,941,966	65,924,625	92.93	4,150,093	5.85	867,248	1.22
Number of employees in service December 31, 1932	300,485	277,180	92.24	18,630	6.20	4,675	1.56

1. Includes only telephone companies reporting to the Interstate Commerce Commission at December 31, 1932.
2. Based on individual balance sheets of all telephone companies. Intercompany duplications included in some of the totals shown in this statement are not readily determinable.
3. Includes only the 69 independent telephone companies operating in the continental United States and reporting to the Interstate Commerce Commission at December 31, 1932.

Source: U.S. Congress, House, *House Report No. 1273* (73rd Cong., 2nd Sess.), "Report on Communications Companies," part 3, no. 1 (1984): 889.

ingly, Splawn recommended a major overhaul of regulation: (a) codification of existing federal legislation in all areas of communications (telephony, telegraphy, and radio and network broadcasting), (b) transfer of jurisdictions from several departments to a new communications commission, and (c) postponement for future action after further study of several controversial subjects.[34]

This study had a major impact, both on its own merits and because Splawn also served as an adviser to the Interdepartmental Committee on Communications created by President Roosevelt in 1933. Commerce Secretary Daniel Roper, who chaired that committee, reported to the president that telephone service, "by its very character," would be "most efficient and satisfactory if conducted as a monopoly." After considering several options, the committee concluded that private ownership, under effective government regulation by a single agency, seemed preferable.[35]

President Roosevelt and Congress accepted this advice without controversy,[36] and legislation prepared by the administration was introduced in February 1934. After a single day of debate in the Senate and a mere two hours in the House, Congress approved the Communications Act of 1934. This law, as one distinguished regulatory scholar has observed, "represents the high-water mark of congressional abdication of power to the regulatory agency."[37] And it stands virtually unchanged today.

Taken at face value, the Communications Act simply consolidated existing federal regulatory authority, as President Roosevelt put it, "for the sake of clarity and effectiveness."[38] By the early 1930s, it was evident that the technologies in telephony, telegraphy, and radio had begun to converge in what might be called a communications industry. Centralized control made sense from an administrative, as well as a reformist, perspective. The act created the Federal Communications Commission (FCC), with seven commissioners, to allocate frequencies, license radio broadcasters, and regulate the rates and facilities of common carriers in interstate "wire communications."

Title II of the act, which applied to telephony, generally "left existing law intact." Yet it provided for "certain additional powers" deemed necessary for more effective implementation.[39] Besides guaranteeing "just and reasonable" rates, the FCC would be responsible for "all charges, practices, classifications, and regulations" (Section 201). Carriers were required to file tariffs, which the FCC could investigate and change "upon its own initiative" (Section 204). Facilities construction would henceforth require a certificate of public convenience (Section 214), and special provision was made for application and approval of consolidations "involving one or more telephone companies" (Section 221). All sorts of transactions, including equipment and service purchases, as well as research, were subject to audit (and possibly future regulation) by the FCC (Section 215).[40]

The most significant change in the Communications Act may have been its statement of purpose. If Congress meant what it said, then national policy was redirected toward a single great social objective:

> [T]o make available, so far as possible, to all the people of the United States, a rapid, efficient, nation-wide, and world-wide wire and radio communications service with adequate facilities at reasonable charges.[41]

If the FCC (and the courts) took this mandate seriously, then the way it allocated costs and set rates could be substantially affected.

Whether or not Congress intended that the *provision* of service to all be centralized would eventually give rise to intense controversy.[42] If monopoly supply were indeed assumed in the act, then the FCC would, by common law, be bound to restrict "duplicative" facilities and competitive entry indefinitely.[43] Although the act itself did not condone monopoly, legislators at the time acknowledged AT&T's monopoly power as they discussed provisions of the bill. "This vast monopoly," read the Senate report, "which so immediately serves the needs of the people in their daily and social lives must be effectively regulated."[44] True to Vail's strategy, Bell System officials promoted this concept of a single monopoly provider. "Telephone is a monopoly," testified President Walter Gifford, "and competition against the public interest."[45]

The commission's first challenge was to figure out how the world's largest firm operated. Indeed, the Splawn report had wisely recommended that several issues be deferred until a detailed study could be undertaken by the FCC. The Telephone Investigation, launched in 1935, is a landmark in regulatory fact-finding. Eighty volumes of staff studies and a 650-page final report were produced during the next four years, documenting in meticulous detail AT&T's organizational structure, its intricate financial history and capital structure, its patent policies, licensing practices, pension system, engineering standards, manufacturing operations, and pricing policies.[46]

The investigation identified a number of regulatory problems peculiar to telecommunications, of which two were procedural: the need for informal regulatory methods, as between the company and the commission, and the need for cooperation with the states.

Because AT&T was so big, because there was no substantive history of formal proceedings, and because the problems were "continuing in nature," the commission staff felt that cooperative methods, through "frank, informal discussion between company and commission representatives," would be preferable to adjudication:

> The atmosphere of the council table seems ordinarily much more conducive to the development of positive results in such matters than does the adversary air which tends to surround most formal proceedings. The aspect of a game or contest which inevitably envelops the respective advocates (be they lawyers, accountants, engineers, or what not) in formal rate cases makes for bickering and bitterness, as well as for delay and expense.[47]

Similarly, the need for cooperation with state commissions appeared self-evident, given the "overlapping and mutual characteristics of most of the problems." In the past, the Bell System's organization and operations had tended to overwhelm state regulatory capabilities, but now, the FCC could help out.[48]

Of the many substantive problems, three were to preoccupy the commission for the next half-century: (a) the problem of separating facilities and allocating costs between state and federal jurisdictions, in order to divide toll revenues between local exchange companies and AT&T Long Lines; (b) rate structure and uniformity across the country, among the various Bell operating companies and AT&T Long Lines; and (c) the need to assess the costs and control the prices of equipment purchased from Western Electric, the Bell System's unregulated manufacturing subsidiary.

Jurisdictional "separations" was the point of departure for federal regulation (see Chart 2-2). If telephone rates for local exchange and interstate long-distance services were to be "just and reasonable" in the public utility tradition, it would be neces-

Chart 2-2
Separations

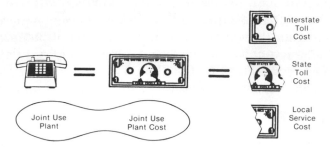

Source: American Telephone & Telegraph Company.

sary to allocate costs and separate the rate base (including the fixed, non-traffic-sensitive plant) between state and federal jurisdictions. But here was the problem: the network itself was inseparable. To make a long-distance call, one used the local exchange plant at each end—telephone, inside wiring, the subscriber line, and local exchange switchboards—as well as AT&T Long Lines' facilities (see Chart 2-3). To price and regulate these

Chart 2-3
Telephone Network Facilities

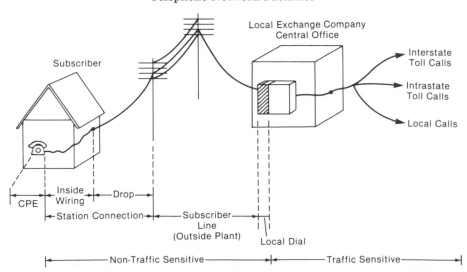

Source: Adapted from John McGarrity, "Implementing Access Charges: Stakeholders and Options" (Cambridge: Program on Information Resources Policy, 1983); © 1983 by the President and Fellows of Harvard College. Used with permission.

elements separately, some rationale, either economic or value-of-service, would have to be found.

The commercial aspects of the problem were already evident by the early 1920s, as interexchange traffic became increasingly important. Local companies had two sources of potential revenues: exchange rates, which were increasingly regulated on a cost basis by state utility commissions, and long-distance charges. The choice was described by AT&T management, in 1924:

> [I]f a number of connecting companies in a given territory are operating with insufficient revenue, it is but natural that in their desire to secure an increase their first thought should be to ask the Bell Company for greater compensation of inter-exchanged traffic. An increase of revenue from this source is not accompanied by a corresponding increase in expense and is therefore so much *velvet* from the standpoint of the connecting company.[49]

For several years, it was the Bell System's policy to maintain the division of revenues on "a liberal basis," to secure "harmonious relations" with independents. Nonetheless, there remained considerable disagreement over what a "liberal basis" should amount to.[50]

Prior to 1930, state regulatory commissions generally adopted a "board-to-board" approach to asset separations. This theory, to which AT&T generally subscribed, held that the cost of local exchange service should be recovered entirely from local rates. Similarly, rates for toll (interexchange) service should only reflect the costs of the toll connection, from one switchboard to the other switchboard. Practicality was the essence of this principle. Although there were occasional discussions of "station-to-station" cost recovery, the difficulties of calculating the necessary separations seemed overwhelming. Moreover, toll service did not seem to impose any significant costs on exchange service per se.[51]

This issue came to a head in 1930, when appeal of a Chicago rate case reached (for the third time) the U.S. Supreme Court. In *Smith v. Illinois Bell,* the appellants claimed that local rates were too high because all the costs of the fixed plant were allocated (board-to-board) to intrastate service. Yet it was an "indisputable fact that the subscriber's station, and the other facilities of the Illinois Company which are used in connecting with the long distance toll board, are employed in the interstate transmission and reception of messages." The Supreme Court more or less agreed, reaching the following conclusion:

> While the difficulty in making an exact apportionment of the property is apparent, and extreme nicety is not required, only reasonable measures being essential . . . it is quite another matter to ignore altogether the actual uses to which the property is put. It is obvious that, unless an apportionment is made, the intrastate service to which the exchange property is allocated will bear an undue burden—to what extent is a matter of controversy.[52]

From this, it seems clear that the Court required a station-to-station jurisdictional separation. But as the last phrase indicates, it offered no guidance to a rational theory of rate making.[53]

Through the war years and on into the early 1950s, state and federal regulators cooperated to develop a satisfactory method of separations. The war-induced boom in long-distance phone traffic pushed AT&T's earnings above the levels that the FCC would accept. As interstate rates were reduced, however, they fell below intrastate toll rates (for similar distances) based on much larger fixed (non-traffic-sensitive) costs. For state regulators, this rate imbalance was politically untenable and lent urgency to the need for resolution. For AT&T's management, station-to-station separations were generally a matter of indifference.[54]

The National Association of Railroad and Utility Commissioners (NARUC) established a task force in 1941 to produce a "separations manual" with AT&T's help. Completed in 1947, the manual adopted accounting methods for station-to-station separations based on "actual use." State by state, non-traffic-sensitive plant actually used to make long-distance calls would be allocated to the interstate jurisdiction in proportion to interstate long-distance *usage*.[55] If interstate long-distance calling accounted for 3 percent of subscriber line usage, then 3 percent of these fixed costs would be allocated to the interstate jurisdiction. It was hoped that this measure would alleviate the FCC's problem of excess profits for AT&T, help equalize toll rates across jurisdictions, and help state regulators hold down local telephone rates. Although the FCC staff participated in the manual's development, the commission declined to endorse it formally.

It became evident almost immediately that these new procedures were still not enough to counterbalance the gains from technological innovation and economies of scale being realized by Long Lines. In fact, by 1950 the problem was getting worse; interstate rates were by then lower than intrastate rates in all but two states. These "inequitable results," reported a NARUC committee, "seem to be due in a very substantial degree to economies resulting from developments in the art of voice transmission, particularly applicable to the longer-haul routes, which are principally interstate." According to AT&T, the cost of Long Lines' circuits had declined from $109 per mile in 1943 to $60 per mile in 1950; the corresponding costs for the Bell operating companies were $134 and $120, respectively.[56]

For the state commissioners meeting in Charleston, South Carolina, the solution lay in adjusting the formula by adding a "subscriber plant factor" to the measure of relative usage. This scheme, which involved a new method of averaging costs for intrastate and interstate plant, would transfer about $200 million from state to interstate jurisdiction (plus about $20 million of variable costs). The FCC balked, however, at the idea of providing what was obviously a subsidy.[57] (This would not be the last time that the commission put economic logic ahead of political sensibility.) Ernest McFarland, chairman of the communications subcommittee of the Senate Commerce Committee, immediately intervened, criticizing the FCC for "shift[ing] the load from the big user to the little user; from the large national corporations which are heavy users of long distance to the average housewife."[58] The commission promptly reopened negotiations with AT&T to reconsider interstate rates, and a compromise approach was found, along the lines NARUC had suggested. When implemented in 1951, this new separations method resulted in a rate *increase,* the first in twenty-five years.[59]

This pattern of shifting the fixed costs of the local exchange plant to recovery from long-distance revenues would continue and expand, as Chart 2-4 indicates, for the next thirty years, eventually creating prices, investments, and patterns of usage that had no bearing whatsoever on costs.

Pricing obviously followed from this basic evolutionary pattern of separations. Nationwide uniform rates initiated during World War I had largely broken down, due to jurisdictional fragmentation and the individual policies of local operating companies, during the 1920s. But once the FCC assumed substan-

Chart 2-4

History of Non-traffic-sensitive Exchange Plant Allocated to Interstate

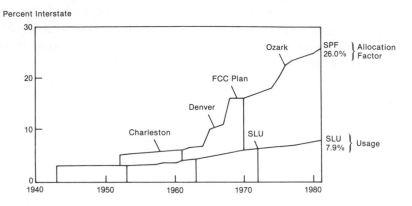

Source: American Telephone & Telegraph Company.

tive responsibility for interstate rates, it became apparent that rate averaging was technically appealing, as well as politically inevitable. It was "universally recognized," said the FCC's general counsel in 1937, "that telephone toll rates cannot successfully be based upon the particular costs involved . . . an averaging process is therefore essential." Moreover, by averaging, "the availability of 'long distance' service to patrons on little-used routes is thus vastly broadened, without unduly burdening any particular group or class of potential customers."[60]

In 1941, the commission adopted a policy of "equal charges for equal services," which the Bell System, however, opposed. For the rest of the decade, the FCC negotiated to reduce, then finally eliminate, interstate rate differentials (for like distances) between Long Lines and the large Bell operating companies that provided interstate service. That, together with the separations subsidy, produced something close to nationwide average pricing by 1952.[61]

Problems associated with AT&T's vertical integration, especially its ownership of Western Electric, posed a third major regulatory issue during this period. Since early in the century, the Bell System operating companies had relied almost exclusively on Western Electric to provide switching, customer premise, and transmission equipment. For the company, this form of vertical integration made eminent strategic sense; it provided economies of manufacturing scale, maintained high-quality engineering standards, and helped enforce uniformity of analog (voice) signals.

For regulators, however, Western Electric was a sort of regulatory black hole in the utility's cost of operations. State and federal commissioners suspected, in the absence of adequate cost accounting, that Western Electric charged exorbitant (i.e., monopolistic) prices for its equipment, prices that the operating companies willingly passed on to customers. Furthermore, this allegedly overpriced equipment inflated the rate base, doubly adding to AT&T's earnings.[62] After producing a massive cost study in 1947, NARUC successfully negotiated price reductions by Western Electric amounting to 13 percent, for a cumulative cost saving of $130 million by 1950.[63]

The anticompetitive aspects of this problem, flagged in both the Splawn report and the Telephone Investigation, also attracted the attention of the Justice Department. Shortly after the war, the staff of the Antitrust Division opened an investigation that built on the data from the Telephone Investigation.[64] In 1949, the Justice Department charged AT&T with violation of the Sherman Act. The government claimed that AT&T had exercised monopoly power through its vertically integrated structure. In particular, the exclusive buying arrangement with Western Electric had produced monopoly profits, injured independent companies through price discrimination, and delayed introduction of more efficient, less expensive types of equipment. The government sought no less than divestiture of Western Electric.[65]

After seven years of litigation, the company submitted to a consent decree. In return for preserving the vertically integrated structure of the Bell System, the company agreed to license its technology freely and to restrict its endeavors to common carrier communications services.[66] By this historic agreement the government reconfirmed the industry structure laid down in the Kingsbury Commitment forty-three years earlier—and end-to-end voice national monopoly. By consent decree, the government attorneys fabricated a Chinese wall between telecommunications and the incipient markets for data processing. Otherwise Bell Labs, at the cutting edge of these developments, could provide AT&T with unfair first-mover advantages.

Ironically, the government and the company had no sooner

settled on preserving the structure of the past than the pressures for change intensified.

Sources of Change

Since the mid-1950s, several factors have interacted to gradually disintegrate AT&T's monopoly and the regulatory regime that supported it. Before examining the process of regulatory reform, we should pause to consider the drivers behind this extraordinary change.

Change in the basic macroeconomic and political context is the most general factor behind deregulation. The inflation and rising interest rates that developed in the late 1960s, followed by stagnating productivity and economic growth in the 1970s, affected many of the underlying economics in telecommunications as well as other regulated industries. As local exchange and capital costs rose more sharply, the growth of usage and revenues slowed. Regulators responded, in part, by increasing the subsidy from long-distance to local service. Politically, the combined effects of Vietnam and Watergate, together with a new consumer movement, severely eroded public confidence in the federal government's authority.

Technological innovation was a second factor. Postwar commercial developments in electronic switching and transmission changed the structural economics of telecommunications in the relative patterns of cost, scale, scope, and entry. These changes in turn affected regulation.

Entrepreneurship was another change driver, induced by a combination of the technological opportunities and economic incentives of regulatory cost allocation. "Political entrepreneurs" like William McGowen or Tom Carter could recognize these peculiar opportunities and follow through with the political action necessary to achieve regulatory approval.

New ideas about the ill effects of regulation and how to regulate were a fourth source of deregulatory change. As early as 1962, academic economists began producing a stream of empirical research that revealed several shortcomings in regulation—especially the rate-base, rate-of-return variety. Meanwhile, economists like Alfred Kahn began promoting new methods of regulation (e.g., marginal cost pricing) and taking their message to Washington.

Finally, regulatory failure was itself an important factor in deregulation. When regulatory institutions, regulatory procedures, or regulators could not adjust effectively to these other changes, their own actions hastened change in public policy. In telecommunications, the FCC's willingness to entertain entry and asymmetrical regulation, despite the continuing growth of cross-subsidies, eventually undermined the traditional structure of their regulation.

Since technological innovation in this industry has been complicated and especially important (although not sufficient) for the process of regulatory change, it is worth exploring in greater detail.

The technological changes in telecommunications that are now apparent to even casual observers originated or were accelerated by defense mobilization during World War II. Large scale research and development efforts in radio and radar technology, electronic circuitry, and materials continued after V-J Day but with a new focus on commercial applications. It was this very refocusing on cost-effective applications and design that precipitated a fundamental breakthrough—the invention of the transistor—and from there, semiconductors, integrated circuits, and microelectronics. The Bell Laboratories were very much at the forefront of these efforts (see Chart 2-5).

During the first postwar decade, the rate of technological discovery outpaced implementation. The first steps entailed expanding transmission and switching capacity, increasing speed, and lowering unit costs by applying a host of *prewar* innovations, including coaxial cable, multiplexing (multiple simultaneous circuits on a single wire), and electromechanical switches. But modernizing the world's largest telephone network was an immense and expensive engineering challenge that could only be accomplished gradually (see Table 2-4).

In the immediate postwar years, voice transmission dominated the demand side of telecommunications. The advent of television, however, created the first new market—long-distance video transmission via high-capacity coaxial cable and microwave. Another new source of demand came more gradually with the development of digital computers, creating a market for data transmission. As the number of computers in use increased from 250 in 1955 to 69,000 by 1968, demand for high-quality, high-speed data transmission (in a digital format,

Chart 2-5
Bell Laboratories Research Focus During World War II

a. Proportion of engineering staff assigned to military

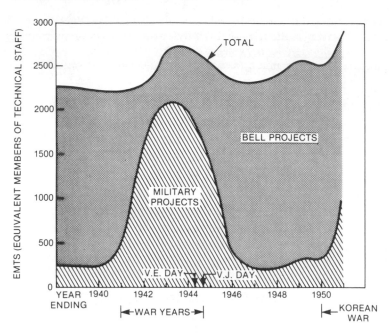

b. Proportion of budget allocated to military

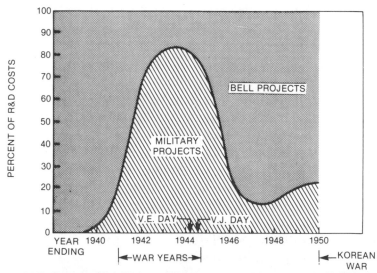

Source: M.D. Fagen, ed., *A History of Engineering and Science in the Bell System,* vol. 2 (Bell Laboratories, 1978), p. 11.

Table 2-4
Postwar Technological Developments in Telecommunications

1946	
February	Coaxial television link, Washington to New York
May	First multichannel, high-frequency microwave system
June	Mobile telephone service in commercial use
October	Two-link microwave radio for television transmission
1947	
November	Completion of transcontinental coaxial cable
December	Transistor invented
1948	
July	First no. 5 crossbar switching system installed
1949	
October	Toll dialing networks on East and West coasts joined
1950	
January	First no. 5 crossbar office for toll switching
March	Step-by-step dial PBX 710E announced
1953	
February	L3 coaxial cable system emplaced
March	Introduction of card translators in toll office
November	Installation of centralized automatic message accounting
1954	
June	Production of color telephones in eight shades
December	Integrated data processing transmitted long distance
1955	
January	Air Defense Nationwide Combat Alert network operative
March	Installation of first electronic switching system
1956	
January	Junction transistor invented at Bell Labs
January	Final judgment in *United States v. Western Electric*

Source: Compiled from AT&T, Historical Archives and Publications Group, *Events in Telecommunications History* (New York: AT&T, 1983).

wherein data were represented as binary digits of electrical pulses) expanded. The network, however, having grown out of a voice technology, operated on analog principles (data represented by the magnitude of electrical signals). Initially, conversion by end-user modems was necessary. By the mid-1970s, the Bell System and some of the independents had begun digitizing their networks, both to lower operating costs and to better serve demand for high-speed data transmission. By the mid-1980s, the network's capabilities for delivering all sorts of communications, including electronic mail, facsimile, telemetry and teletype, voice, and video, were significantly enhanced and integrated.

Meanwhile, technological innovations on the supply side affected both the links (transmission) and the nodes (switches and customer premise equipment) of the network. As the relative efficiency gains (and thus cost reductions) in switching exceeded those in transmission, the network architecture (hence industry structure) began to change.[67]

Before World War I, switching technology basically consisted of human operators sitting in front of large plug boards and manually moving around connector plugs. During the next several decades, electromechanical switching systems helped ease this high-cost bottleneck. After World War II, these systems were made by means of relay springs.[68] Although AT&T installed its first electronic switching system in 1955, the thousandth such installation (out of 9,800 BOC local exchanges) did not occur until 1976. By 1985, about half of all calls were digitally switched.[69]

While application of electronic switching increased scale and drove down costs in central telephone exchanges, it also made possible smaller packaging, and thereby dispersion, of switching intelligence. A key device called a private branch exchange (PBX) was an end-user switch (essentially a small computer) serving anywhere from 10 to more than 5,000 lines. By 1985, the network included about 50,000 PBXs (of which 14,000 served more than 400 lines apiece).[70]

In long-distance transmission technology the innovations were almost as impressive. Steady improvements were made in coaxial cable (a copper tube with wire running down its center) and multiplexing, to the point where, by the late 1970s, trunk cables could carry more than 100,000 channels. Joining this technology in the late 1940s was microwave radio, a direct result of wartime radar development. Microwave relays, employing very high frequencies, made it possible to transmit communications signals along a narrow path through the atmosphere, between line-of-sight antennas atop towers. Early systems, which carried up to a thousand voice circuits, already accounted for about a quarter of AT&T's long-distance network by 1959.[71] With multiplexing and other enhancements, these systems have recently achieved capacities in the range of 35,000 circuits.[72] Geostationary satellites, placed in orbit during the mid-1970s, provided a third means of transmission, especially suited to very long distances. And finally, the recent commercialization of fiber optic cable, through which communications signals are carried by lasers, is causing a dramatic reduction in high-volume unit transmission costs, as well as a massive increase in transmission capacity.

Transmission technologies, as Charts 2-6a and 2-6b indicate, contributed to substantial reductions in the fixed costs of long-distance service. Furthermore, the distinctive characteristics of each new technology created opportunities for specialized service offerings. Satellites, for example, were well suited to transoceanic service and point-to-multipoint applications, such as television, private networks, and value-added networks. Microwave, by contrast, was ideal for point-to-point service, either in rugged terrain or between pairs of large urban markets. Since the capital costs and right-of-way costs of microwave were small relative to cable, as was minimum efficient scale, this technology potentially lowered entry barriers to the interexchange and especially private line (for business customers) markets.

The technologies for electronic switching not only reduced the cost of end-user equipment and central office switching but undermined the economies of manufacturing scale. These changes in cost characteristics lowered the entry barriers to certain segments of the equipment market. Historically, Western Electric's dominance was partially based on scale economies associated with manufacturing electromechanical devices. Whereas the telephones and central office switches in which Western Electric specialized were used exclusively for telecommunications, electronic components were produced and used for military applications, consumer electronics, and computers For these electronic components Western Electric had no significant scale advantages compared with a dozen other manufacturers.[73] By the mid-1960s, it was possible for a relatively small business to assemble a telephone or PBX from widely available electronic components, at costs competitive with those of Western Electric. By offering new features that microelectronics made possible and by focusing on the business market (thus minimizing marketing and distributional barriers), an entrepreneur could not only find a niche but also scarcely avoid profitability.[74]

The economics of these technological changes, especially *in combination,* posed serious problems for the regulatory status quo. For example, the use of identical electronic components by "computer" and "telecommunications" manufacturers was only

Chart 2-6a

Investment Costs for New Terrestrial Transmission Systems

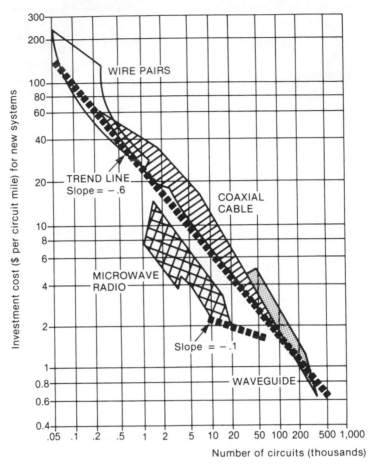

Source: Charles River Associates, redrawn from E. O'Neill, "Radio and Long-Haul Transmission," *Bell Laboratories Record,* January 1975, p. 55.

one facet of a larger trend toward convergence. According to an engineering executive with Advance Micro Devices Inc., "It is the computer control of the telephone switching machines, and the telephone network capability to switch and transmit computer signals that is causing the two fields to grow together."[75] In the process, substitution, cross-entry, and competition put greater and greater pressure on the 1956 consent decree, forcing the FCC to implement detailed rules.

Similarly, the system of cross-subsidies so recently estab-

Chart 2-6b

AT&T Long Lines Average Annual Investment per Circuit Mile, 1940–1975

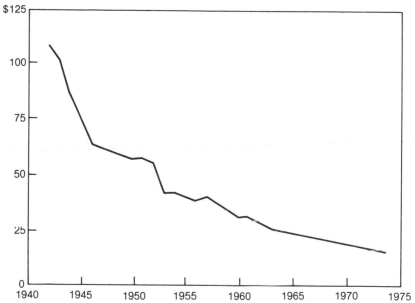

Source: AT&T (1977), in L. Johnson, *Competition and Cross-subsidization* (Rand, 1982), p. 45.

lished to equalize prices between local and long-distance services came under increasing pressure. As Chart 2-7 shows, the new technologies in transmission and switching dramatically lowered the costs of long-haul interexchange services but scarcely slowed the rising costs of the local (exchange) loop. Unless reversed, the federal-state pricing compromise could only exacerbate the economic distortions already created. Herein lay the seeds of regulatory failure.

At the broadest and most complex level of aggregation, these innovations were gradually changing the architecture of the network. Until the 1960s, the interexchange system had been organized by Bell engineers as a five-level hierarchy of central office switches. The technology and its economics made this the most efficient possible design.[76] But as distributed processing became more economical, through microelectronic advances in PBXs and microcomputers, this hierarchy became somewhat less centralized and more permeable. In retrospect, regulatory barriers to entry appeared less sustainable, while substitution

Chart 2-7

Changing Cost Characteristics in Telecommunications

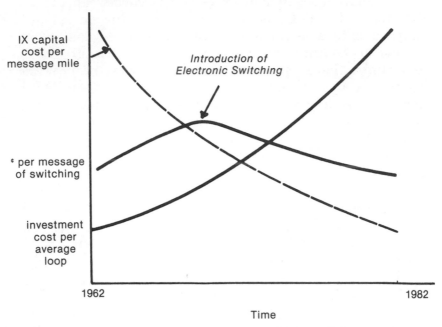

Source: Bell Laboratories, reprinted in R. Vietor and D. Dyer, *Telecommunications in Transition* (Boston: HBS Case Services, 1986), p. 20.

and horizontal fragmentation seemed likely in the face of ever larger pricing distortions.

The political process of regulation, despite its immense inertia, was bound to change eventually.

The Process of Deregulation

The process of regulatory change in telecommunications has been exceedingly gradual and, as it is still, ad hoc. Separate threads of change ran through each major market segment: customer premise equipment (CPE), interexchange long-distance service, enhanced services, and, eventually, local exchange service. Procedurally, the process was immensely complicated, with regulatory cases dragging on for years and years. As the process gained momentum, new competition put increasing pressure on ancillary regulation, and the process became intensely politicized.

Customer Premise Equipment

Deregulation began more or less with a rubber cup. For years, the Hush-a-Phone Corporation had been selling a device that snapped onto a telephone handset to form a cuplike shield around the transmitter, intended to block out background noise and muffle the user's voice. The device made no physical or electrical contact with the circuit whatsoever; it was just a rubber cup. Yet in 1948, the Bell operating companies informed Hush-a-Phone and many of its customers that use of the device was contrary to the restriction on "foreign attachments" that were written into AT&T's state and federal regulatory tariffs. These restrictions had originated decades earlier, as a technological by-product of AT&T's end-to-end service responsibility and its strategy.[77]

Hush-a-Phone complained to the FCC, seeking to qualify the blanket prohibition on foreign attachments in AT&T's interstate tariffs. It took the commission *seven years* to decide the case. In 1955, the commission upheld AT&T's position on the grounds that "unrestricted use of the Hush-a-Phone could result in a general deterioration of the quality of interstate and foreign telephone service."[78] Not surprisingly, Hush-a-Phone appealed the decision. In 1956, the same year as the consent decree in *United States v. Western Electric,* the court of appeals set aside the FCC's order, applying a standard of "private benefit without public detriment."[79] AT&T subsequently revised its tariffs but did so in the narrowest possible manner. Customer-provided equipment that entailed interconnection by electrical contact remained totally prohibited, while other devices would have to be proven to be nonharmful.

The real test on entry into the market for CPE came more than a decade later, in the *Carterfone* case. Between 1959 and 1966, the Carter Electronics Corporation of Texas manufactured and sold a device that permitted direct voice communications between a person using the telephone network and a person with a mobile radio. A telephone receiver, placed on a cradle in the Carterfone, effected an inductive (but not electrical) connection between the telephone line and the mobile radio channel. This arrangement fulfilled a market demand for remote telecommunications on the part of cattle ranchers, oil field workers, and offshore drilling platforms.

Whenever Bell employees discovered a subscriber using a Carterfone, they would discontinue service, in compliance with the foreign attachment restrictions in their tariffs. Tom Carter, the president of Carter Electronics, would promptly give a refund to the subscriber, reclaim his Carterfone, and resell it. In this manner Carter managed to serve 10,000 subscribers with just 3,500 devices.[80] In 1965, after negotiating unsuccessfully with AT&T and complaining informally to the FCC, Carter filed an antitrust suit seeking an injunction and treble damages. The court referred the case to the FCC on the basis of primary jurisdiction.[81]

The FCC considered the need for the Carterfone, its effect on the telephone network, the applicability of AT&T's tariff restrictions, and whether or not such restrictions were generally "just and reasonable." This last point obviously elevated the case to landmark proportions. In its decision, the FCC held that "the tariff has been unreasonable, discriminatory, and unlawful" and that "provisions prohibiting the use of customer-provided interconnecting devices should accordingly be stricken."[82]

The FCC decision opened wide the CPE market to entry. With telephones, decorator phones, recorders, chimes, headsets, key sets, PBXs, teleprinters, and modems, entrepreneurs entered the market. On they came: Multiphone, Phone-Mate, Code-a-Phone, Attache Phone, Selectron, Tele/Resources, Wren, Codex, Vadic, and a score of others.

Despite the commission's suggestion that AT&T develop "reasonable standards" to prevent the use of harmful devices, the company chose to revise its tariffs so as to require use of a "protective device" for any non-Bell equipment. These devices, which the Justice Department subsequently alleged to be over-engineered and excessively expensive, had to be installed by Bell operating companies on a leased basis, regardless of any threat of harm.[83] Although the FCC and the National Academy of Sciences endorsed the reasonableness of these procedures, they remained a source of regulatory controversy and litigation for another decade.[84]

Under constant pressure from non-Bell equipment suppliers, the FCC initiated a proceeding to assess the problems of interconnect. Three years later, in 1972, the commission adopted a program of certification based on detailed technical standards in lieu of protective devices. AT&T and several other telephone companies appealed the program, but the court upheld the FCC in 1977. Although AT&T made an effort to require customers to take at least one telephone per line from the telephone company, this too was rejected by the FCC. Certification became fully operative in 1978.[85]

Enhanced Services

As innovation, new entry, and competition reshaped equipment markets, it became increasingly difficult to distinguish the communications functions from the data processing functions that the equipment provided. This blurring of product-market boundaries posed a dilemma for common carrier regulation and antitrust enforcement. For AT&T, constrained by the 1956 consent decree, the blurring gave rise to incredibly complicated strategic and organizational problems.

In its First Computer Inquiry, begun in 1966, the FCC addressed two fundamental questions: (a) "whether data processing services should be subject to regulation," and (b) "whether common carriers should be permitted to engage in data processing."[86] Given the "state of the art [in computer and communications technologies] as it then existed," the commission settled on a definitional standard based on a criterion of relative use. It would not regulate pure data processing or hybrid services where message switching was an incidental feature of an integrated data processing service. It would, however, regulate hybrid services in which data processing was incidental to message-switching functions. The commission adopted a policy of "maximum separation," whereby a common carrier could provide data processing services only through a separate corporate entity.[87]

When the FCC adopted these rules in 1971, their workability depended on the general separation of mainframe computers and central office switches. But microelectronics technology was simply moving too fast. Almost immediately, the FCC realized that "the distributed processing environment, wherein computer processing capabilities are placed throughout a data information or transmission system," had combined with consumer demand to "overrun the definitions and regulatory scheme of the *First Computer Inquiry*."[88] And so a Second Computer Inquiry was begun.

In this proceeding, which continued throughout the remainder of the 1970s, the commission and its professional staff made a valiant effort to fulfill their statutory obligations in a manner consistent with the technological and economic imperatives. For network services, the commission adopted a new and simplified definition that distinguished between "basic" and "enhanced." Basic service was limited to the common carrier offering of transmission capacity that *moved* information; enhanced service combined basic service with processing applications that *acted upon* the format or content of the information. AT&T and GTE (General Telephone) could offer enhanced services but only through separate subsidiaries.

This dichotomy, however, did not resolve the problem of separability between functions and equipment. Initially, the commission tried to distinguish among kinds of CPE on the basis of processing functions. But every classification proposal appeared arbitrary or obsolete. The conclusion was inevitable: no distinction could be made, and so all CPE should be accorded "uniform regulatory treatment." With this realization, the focus shifted to the role of communications common carriers in providing CPE. The commission now followed its own logic to the next step: deregulation and detariffing of equipment altogether.[89] The landmark decision, known as Computer Inquiry II, was adopted April 7, 1980. Thirty-two years had passed since Hush-a-Phone filed its case.

Interexchange (Long-distance) Service

In 1956, the same year it decided the Hush-a-Phone case, the FCC opened a proceeding to consider allocation of radio frequency bands above 890 megacycles. The new technology of point-to-point microwave transmission was ideally suited to this range. Allocation of the remaining unused capacity therefore attracted considerable interest. The salient issues were both technical and economic:

1. What were the prospective supply and demand for these frequencies, and would private use cause harmful interference or terminal congestion?
2. If private point-to-point use were warranted, could private systems be shared? And who should be eligible?

3. What effects would private point-to-point systems have on common carrier services, where both are available?
4. Is the FCC obliged to protect users of common carrier service from any adverse economic effects that the carriers might suffer from the operation of point-to-point systems?
5. To what extent would private point-to-point communications systems depend on interconnection with common carriers?
6. Would a policy of restricting private point-to-point systems "result in a lessening of competition or a fostering of monopoly in the manufacture, sale, use, or provision of communications facilities contrary to the public interest?"[90]

The parties interested in these issues included communications common carriers, independent manufacturers of microwave equipment (RCA, Motorola, and Collins Radio), right-of-way companies (utilities, pipelines, and railroads), government agencies (supplying police, fire, highway, and forestry services), and a variety of potential vendors and users of specialized services (e.g., trucking, mining, construction, agriculture).

The technical issues were most easily resolved: improvements in multiplexing and transmission technology were expanding capacity and alleviating interference far more rapidly than demand was growing.[91] On economic issues, however, the common carriers were unanimously opposed to licensing private systems for anything other than public safety and right-of-way use. They argued that private systems were redundant and thus wasteful, that they posed problems of interference, and that they detracted from the spread of universal service. AT&T in particular focused its opposition on the threat of "cream skimming"—the threat that private systems could skim off the most profitable routes, leaving less profitable traffic to the common carrier, an eventuality that could only result in higher rates for the public network. AT&T also raised the issue of uncontrolled interconnection, which would cause physical, as well as economic, harm to the network. And most of all, AT&T opposed any sharing of private systems "because it would result in an aggravated form of cream skimming, and would 'snowball' to unreasonable proportions."[92]

At the time, however, these concerns seemed remote. Potential entrants and suppliers clamored for a competitive, open-entry policy. The market, they argued, was tiny, had nothing to do with the public switched network, and could not possibly harm the giant Bell System. The Justice Department opined that in the Communications Act, congressional intent relied "upon competition to regulate and develop the communications field to the greatest extent possible."[93]

In 1959, the FCC announced its decision in *Above 890*. Licensing of private microwave systems would be substantially liberalized. The need and the opportunities were obvious, while the dangers merely imputed. The commission viewed private microwave as a technological opportunity to stimulate diversity and expand the market for telecommunications services; any traffic losses to common carriers would be inconsequential. On the complicated issue of interconnection, the FCC waffled: "generally speaking, these are matters . . . which are required to meet the statutory tests of justness and reasonableness." The commission rejected cooperative arrangements at that time, since it did "have some concern" that cooperative facilities could "have many of the attributes of communication common carriers without assuming the responsibilities of service."[94] Indeed.

Despite some aggressive efforts by AT&T (discussed elsewhere in this chapter) to hold customers and block entry into the private line market, separations and national average pricing made low-cost, high-volume routes an attractive target. As more and more large corporations began developing private networks for intrafirm voice and data communications, at least one entrepreneur got the idea of reselling microwave channels in smaller lots to smaller customers. In December 1963, John Geoken, the owner of a radio repair business in Joliet, Illinois, created Microwave Communications, Inc. and applied to the FCC for authority to construct and operate a "limited, common carrier microwave service" between St. Louis and Chicago.

Although MCI (as it was later renamed) offered no new technology or communication functions, it did propose a new product—discount private line service. MCI's rate (for two voice-grade channels, St. Louis to Chicago) would be $464 monthly, and each channel could be shared by five users; for higher-quality circuits, AT&T charged $988 and prohibited sharing. MCI's customers, moreover, would have to provide their own means of interconnecting with MCI's transmitters. The entire system was initially estimated to cost $564,000.[95]

AT&T (and several independent telephone companies) op-posed MCI's application on the grounds that the service was unwarranted and unreliable and that the effect was nothing more than cream skimming. But at the FCC, the staff of the Common Carrier Bureau, the hearing examiner, and a majority of the commissioners did not see it that way. During six years of filings, revisions, appeals, and interventions "the struggle be-tween little MCI and the giant carriers" took on a "David and Goliath" character.[96] The company barely survived, thanks only to a cash infusion and new management from Bill McGowen, a wily venture capitalist.[97]

On a four-to-three vote in 1969, the FCC approved MCI's ap-plication. Long-distance private line service was open to compe-tition, presumably isolated by FCC tariffs from the public switched network of the Bell System. Commissioner Nicholas Johnson, concurring with the majority, candidly expressed the dissatisfaction with monopoly and regulation that had begun to be felt in Washington:

> On this occasion three Commissioners are urging a perpetuation of more Government regulation of business, and four want to experiment with the market forces of American free private enterprise competition as an alter-native to regulation.
> No one has ever suggested that Government regulation is a panacea for men's ills. It is a last resort; a patchwork remedy for the failings and special cases of the marketplace. . . . I am not satisfied with the job the FCC has been doing. And I am still looking, at this juncture, for ways to add a little salt and pepper of competition to the rather tasteless stew of regulatory protection that this Commission and Bell have cooked up.[98]

But for the three dissenting commissioners, this "experiment" seemed the height of folly. "The decision of the majority," wrote Chairman Rosel Hyde, "is diametrically opposed to sound eco-nomics and regulatory principles. It likewise is designed to cost the average American rate payer money to the immediate benefit of a few with special interests." It was indeed cream skimming that would "destroy the principle of nationwide aver-age ratemaking."[99]

After the MCI decision, the floodgates opened. During the next few months, thirty-three applicants filed 46 separate pro-posals for more than 1,877 microwave stations. Among these were 17 MCI affiliates; Southern Pacific, proposing a 95-station network across the southwest; Western Tele-Communications, blanketing the far west; and Datran, which planned to build a nationwide digital/data network of 214 stations.[100] Entry, observed AT&T caustically, was no longer "experimental." This wave of proposals, if granted, "would seriously undermine the policy of uniform interstate rates and dilute or delay the benefits that economies of scale would otherwise make available to the general telephone-using public."[101]

The commission responded with a generic proceeding, for *Specialized Common Carrier Services,* to address the two central issues. First, would the public interest "be served by permitting the entry of new carriers in the specialized communications field?" The FCC staff viewed "specialized communications," which included private line microwave, satellite relay, and digital/data transmission, as a separate *market,* rather than a separate *tariff.* And second, if entry were unrestricted, what was "the appropriate means for local distribution of the proposed services?" In other words, should interconnection with local exchange, telephone operating companies, in order to provide the critical "last mile" of access to customers' premises, be obligatory? Without access, as McGowen would soon attest, specialized carriers were all but impotent.

The FCC staff presented a host of arguments in favor of open entry to this "specialized market": demand for these "diverse and flexible" services was growing rapidly, economies of scale had become less important, and competition would benefit equipment manufacturers and serve as a yardstick for efficient pricing. The threat of cream skimming was allegedly trivial, since only 2 to 4 percent of AT&T's services were vulnerable.

The commission concluded that "all but a very few of the parties commented in support of the staff analysis." This meant AT&T, GTE, United Telephone, United States Independent Telephone Association, and NARUC, who were all opposed. AT&T, sensing a losing battle, urged that there should, at least, be "no protection of inefficient specialized carriers who have no economic basis for survival." To be fair, AT&T should be allowed to deaverage rates and use long-run incremental costs as the basis for competitive pricing.[102]

Midway through 1971, the FCC issued its ruling. First, open entry and competition would be allowed for specialized services. Second, the Bell System (and other local exchange carriers) would be required to provide interconnection with any new carrier "on reasonable terms and conditions," as well as "local distribution services" for its customers. And third, the FCC saw no good reason to depart from the historic approach to pricing and regulation for AT&T, since that approach "has been generally regarded as consistent with the public interest."[103]

The reduction of regulatory barriers to entry, however, did not guarantee success for MCI. The company continued to struggle with its financial and regulatory problems. In January 1972, after more than a year of difficult negotiations with the Bell System to arrange for interconnection, MCI commenced operation. After boldly announcing plans to build a 100-city network, the company made a stock offering of $33 million and secured a credit line for $64 million. But as operating losses continued throughout 1972 and 1973, MCI's expansion plans were cut back to 41 cities, and the focus on bare bones, point-to-point private line service was reassessed. Early in 1973, Bill McGowen evidently began redirecting his company's strategy.[104]

The fight over FX Service that began in early 1973 between MCI and AT&T was an extraordinary episode in the strategic use of regulation. It illustrates the very essence of government intervention in markets. FX (for "foreign exchange") service was (and still is) a premium service offered by AT&T; it provides first a dedicated private line from a customer's premises to an interexchange switchboard elsewhere in the Bell System (rather than on other premises) and then a local line for local service. A business customer in New York, for example, could make calls throughout metropolitan Boston *as if* they were local calls. MCI and several of its customers wanted this service.[105]

In February 1973, MCI sent a letter to AT&T, accusing it of bad faith in interconnect negotiations and demanding FX connections. Kenneth Cox, MCI's senior vice president who had recently resigned as FCC commissioner, sent a blind copy of the letter to the chairman of the FCC, with a note that said, "I would greatly appreciate your reading this so that you can get some feel for the difficulties we are having in our dealings with AT&T."[106] Shortly thereafter, McGowen met with John deButts, AT&T's chairman, and threatened an antitrust suit if AT&T would not provide better access (i.e., FX service) to its local operating companies. DeButts, who viewed FX and similar services as joint offerings that AT&T was not obligated to provide, refused.

During the next six months, MCI lobbied the FCC to clarify

that MCI's interconnection rights with AT&T included FX service. The central figure in these machinations was Bernard Strassburg, chief of the Common Carrier Bureau and a leading advocate of competition within the FCC since 1963. In early October, Strassburg met with MCI's Ken Cox to discuss the FX situation. That afternoon, Strassburg drafted an order involving interconnect tariffs and placed it before the commissioners, as a routine matter, for their signature. That order subsequently obligated the Bell System to provide FX connections to MCI.[107]

The endgame was not long in coming. MCI expanded its network of private line and FX services throughout 1974. In September it filed changes in its tariffs, to include a new Execunet service, and placed a large number of orders with AT&T for local exchange connections. Although FCC staff requested clarification of these tariff amendments, MCI's plan was revealed only when its promotional material became available in 1975. Using any touch-tone phone, an Execunet customer could dial MCI's local number, provide his or her billing number to gain access to MCI's long-distance circuits, and then dial directly any telephone in another city where MCI, through an FX connection, offered Execunet.

MCI, through the magic of FX connections between major cities, had secretly violated the government's artificial barrier and entered the sacrosanct market for switched long-distance telephone service. Execunet competed directly with AT&T's Message Toll Service (MTS) and indirectly with its Wide Area Toll Service (WATS). (Chart 2-8 illustrates the different basic services.) Discovery caused consternation at AT&T and the FCC.[108]

To the commissioners, Execunet was "a breach of faith."[109] Finally the FCC realized the implications of its policy of creeping, specialized entry. Now it tried to step back, unanimously ordering MCI to cease offering Execunet.[110] But it was too late. McGowen immediately challenged the order in court, and won. The court ruled that the commission had failed, in *Specialized Common Carriers,* to impose specific restrictions on private line connections. The commission, now squarely behind AT&T, tried another approach in 1978. It ruled (in *Execunet II*) that the Bell System was not obligated to interconnect with carriers providing the functional equivalent of MTS service. McGowen appealed again, and won.[111] The long-distance telecommunica-

Chart 2-8
The Structure of Basic Telephone Services

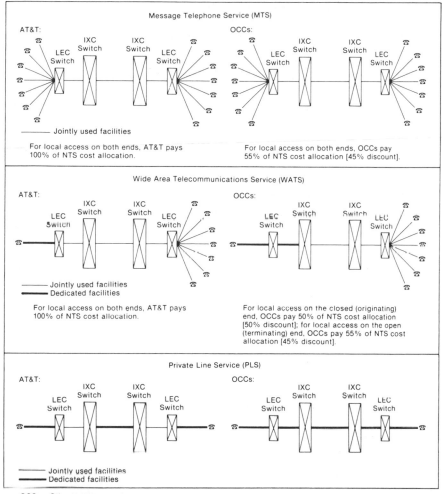

Source: Adapted from Vietor and Dyer, *Telecommunications in Transition,* p. 103.

tions market, after one hundred years of monopoly, was now open to competition.

The suddenness of this development posed some immediate regulatory problems that required solutions. The court, in ordering interconnect of local facilities, did not address the matter of compensation (for originating and terminating long-distance calls). Prior to Execunet, as we have seen, that matter was accomplished through the elaborate process of separations and division of revenue.

MCI and other specialized common carriers had interconnect tariffs that reasonably reflected the fully distributed cost of connecting a private line. These rates bore none of the additional cost burden ("subscriber plant factor") that had gradually been shifted to AT&T Long Lines over the years. Immediately after the court's decision, the Bell operating companies filed new tariffs to recover as much per call from MCI and other new entrants as they received from AT&T Long Lines (which was more than three times as much). Not surprisingly, state regulators backed the Bell companies to avoid losing long-distance revenues that subsidized local service.

MCI and the other newcomers adamantly objected on the grounds that the quality of their local interconnections was inferior to AT&T's (e.g., more noise in the connection, the need to use touch-tone phones, and 22-digit dialing by their customers). They also claimed that the proposed charges would bankrupt them. Under pressure from interested congressmen, and in the hope of achieving an interim solution that might provide "rough justice," all parties agreed to negotiations, and a compromise was reached—the Exchange Facilities for Interstate Access (ENFIA) tariff.[112] Under ENFIA, the new long-distance carriers would bear some of the local subsidy but at a significant discount from AT&T: initially, 65 percent less (based on estimated minutes of local usage) and gradually shrinking (as their combined revenues increased) to only 45 percent less.

In the meantime, the FCC would thoroughly reexamine the entire structure of telecommunications markets, as well as the historic system of separations and division of revenues that had made universal service a reality.

Market Structure in Transition

By 1979, the nationwide end-to-end voice monopoly that public policy had confirmed in 1956 was a shambles. Although it had remained intact until the late 1960s, thereafter it disintegrated rapidly.

In the terminal equipment sector, the boundary between switching equipment and data processing was virtually gone (see Table 2-5a). Computer manufacturers were rapidly gaining a foothold in PBXs and modems while AT&T was reorganizing to enter the minicomputer and microprocessing markets through a separate subsidiary. AT&T's market share had eroded dramatically in several product lines. In the critical PBX market, where Northern Telecom, Rolm, Mitel, and NEC were pushing hard, AT&T's market share fell to 29 percent by 1982.

In the interexchange sector (see Table 2-5b), AT&T was still overwhelmingly dominant, but the rate of entry was almost unbelievable. Besides more than a dozen facilities-based entrants, like MCI and Southern Pacific Sprint, several hundred resellers were setting up to arbitrate the difference between AT&T's wholesale (WATS) and retail (MTS) services.

Private networks, capable of providing a variety of enhanced services, were catching on among large users. Meanwhile, all sorts of entrepreneurs were gearing up to offer such services competitively. Besides offering value-added services, these new systems "bypassed" the public switched network, effectively avoiding the high rates for access that subsidized local service.

Only in the local exchange markets, where the Bell System had an 82 percent share, did monopoly still prevail. And even this had been called into question by the Justice Department in a criminal antitrust suit that sought divestiture.

AT&T's Strategic Response

The result of AT&T's response to the creeping onset of competition was ironic. Nearly every step its management took, from regulatory tactics to product policy, was subsequently numbered among the antitrust charges that eventually lead to dissolution.

Looking back at this course of action, one wonders how different decisions by AT&T might have affected market structure and the progression of regulatory change. Certainly AT&T's management made some important decisions, especially between 1970 and 1982. But given the technological drivers, the immense economic opportunities for determined entrepreneurs, and the ideological presumptions that big is bad and competition is good, AT&T could scarcely avoid stepping on competitors' toes, if it were to respond at all.

In the earliest stages of regulatory transition—*Hush-a-Phone*

Table 2-5a

Terminal Equipment—Thousands of Units Sold

	1970	1975	1978	1979
PBX (thousands of lines):				
Western Electric	900	950	1,150	1,300
GTE-Automatic Electric	65	60	70	80
Unaffiliated United States	89	484	829	989
Foreign	66	109	121	140
Facsimile:				
Western Electric	—	—	—	—
GTE-Automatic Electric	—	—	—	—
Unaffiliated United States	10	15	36	44
Foreign	0.5	1	3	3
Modems:				
Western Electric	40	151	205	223
GTE-Automatic Electric	5	10	16	20
Unaffiliated United States	38	140	205	217
Foreign	—	—	—	—
Key telephones:				
Western Electric	1,231	1,429	1,750	1,800
GTE-Automatic Electric	90	116	123	124
Unaffiliated United States	22	75	94	96
Foreign	21	84	123	132
Decorator telephones:				
Western Electric	—	200	300	450
GTE-Automatic Electric	—	—	—	—
Unaffiliated United States	50	332	514	668
Foreign	20	8	59	70
Answering machines:				
Western Electric	—	—	—	—
GTE-Automatic Electric	—	—	—	—
Unaffiliated United States	30	43	70	75
Foreign	115	110	140	170
Automatic dialers:				
Western Electric	—	40	70	80
GTE-Automatic Electric	—	—	—	—
Unaffiliated United States	30	13	38	51
Foreign	50	62	140	178

Source: U.S. Congress, House, Committee on Energy and Commerce, *Telecommunication in Transition: The Status of Competition in the Telecommunications Industry* (97th Cong., 1st Sess.), Committee Print, November 1981, pp. 124, 184–85.

Table 2-5b

Intercity Transmission Revenues

[Dollar amounts in thousands]

Service	1979		1980	
	Revenues	Share percent	Revenues	Share percent
Switched voice (Mts. Wats, equivalents):				
Joint through service:				
AT&T	$21,840	73.6	$24,292	72.6
Independents	4,809	16.2	5,494	16.4
Totals JTS	26,649	89.8	29,786	89.0
Specialized common carriers:				
MCI	76	.3	153	.5
SPCC	58	.2	129	.4
USTS	3	.01	19	.06
Others	138	.5	305	.9
Private line service:				
Joint through service:				
AT&T	1,972	6.7	2,292	6.9
Independents	121	.4	130	.4
Totals JTS	2,093	7.1	2,422	7.3
Specialized common carriers:				
MCI	44	.1	51	.2
SPCC	36	.1	35	.1
USTS	17	.06	22	.07
Total SCC	100	.3	114	.3
Western Union	138	.5	168	.5
Record message service:				
Western Union	312	1.1	350	1.0
International record carriers	10	.03	9	.03
Other services:				
Miscellaneous common carriers	32	.1	41	.1
American Satellite	16	.05	19	.06
RCA American	44	.1	64	.2
Total	$29,671	100.00	33,448	100.00

Source: U.S. Congress, House, Committee on Energy and Commerce, *Telecommunication in Transition: The Status of Competition in the Telecommunications Industry* (97th Cong., 1st Sess.), Committee Print, November 1981, pp. 124, 184–85.

and *Above 890*—one would hardly expect a significant organizational or political response from so large and self-confident an institution as the Bell System. These cases just scratched the far periphery of the monopoly communications market. From the perspective of 195 Broadway (AT&T's corporate offices), those cases were merely two of several dozen federal and state regulatory dockets in which the company was continuously involved. All sorts of other dockets, involving separations, tariffs, and new facilities, shared the attentions of AT&T's Regulatory Affairs Department at the time. Given decades of experience with detailed and time-consuming regulatory procedures, AT&T devoted considerable resources to participating carefully and patiently in every docket touching on the network.

But AT&T's responses were not limited to the regulatory arena. TELPAK was an aggressively priced product response to entry in the private line segment that fairly illustrates AT&T's competitive and strategic dilemma. Immediately after the FCC decided *Above 890,* the Rate Planning Group at AT&T began studying the prospects for a broad-band (voice, video, and data) service offering that would compete with the capabilities of private microwave networks (AT&T offered only individual circuits at the time). Although the market was small (estimated at $62 million), its potential for growth was immense. In 1960 the group completed its study, *The Broad-band Report,* recommending TELPAK, a private line, bulk-rate offering. The problem, according to the report, was value-of-service pricing, not the Bell System's cost structure or a lack of technology. Because of price averaging and separations, AT&T's high prices for business services were unrelated to their costs. Open, selective entry into the private line market made this situation unsustainable.[113]

The Broad-band Report was discussed and adopted at the AT&T President's Conference in November 1960. Shortly thereafter, the company filed its TELPAK tariff, with four classifications: TELPAK A, B, C, and D, providing bundles of 12, 24, 60, and 240 private line channels, respectively. The new rates offered a huge discount over existing rates, yet they were nonetheless justified by costs, according to AT&T. The rate for TELPAK A, for example, was $1,860 for 12 channels, versus $3,780 at prevailing private line rates. The rate for TELPAK D, for the largest users, was $11,700, versus $75,600 at prevailing rates—a discount of about 85 percent.[114]

Motorola and Western Union immediately challenged TEL-PAK, and the FCC instituted an investigation. A host of large corporate users that would benefit substantially from the lower rates intervened in support of TELPAK. In 1964, the FCC ruled against TELPAK A and B on the grounds that competitive necessity did not warrant the degree of discrimination between these small bundles and "like services" (e.g., the individual private line rate).[115] For TELPAK C and D, the commission felt there might be a competitive justification, but it could not determine costs. The FCC ordered AT&T to prepare a cost study for TELPAK C and D as part of a larger effort ("The Seven-Way Cost Study") to separate the incremental costs of AT&T's monopoly and competitive services. In the context of this study and with multiple appeals, revisions, and reconsiderations, the TEL-PAK case lasted fifteen years. In 1976 the commission reversed its earlier conclusion and ruled that TELPAK C and D were unjustified by competitive necessity and thus unlawful.[116] Yet all the while, AT&T had been offering the service. The Justice Department alleged that TELPAK was a prime example of AT&T's monopolistic behavior.

It was the FCC's decisions in *Carterfone* and *MCI* together that triggered a fundamental reassessment of strategy by AT&T's senior management. H.I. Rommes, a thoughtful, reserved engineer who had previously headed Western Electric, was chief executive officer of AT&T at the time. He had not led actively (and perhaps did not need to) until the spring of 1970. In the February annual report, he first called for "thinking through" the role of competition. "Where competition would benefit the public," said Rommes, "competition should be encouraged."[117] Two months later he established an Executive Policy Committee (EPC) to more formally and explicitly address the strategic implications of changing public policy.[118]

In his remaining year and a half as chairman (and while a progressive illness took its toll), Rommes and his committee got no further than articulating the company's goal. AT&T would compete in peripheral markets, but the focus would remain on the public switched network and on improving its capabilities.[119]

John deButts, AT&T's vice chairman and a member of the EPC, succeeded Rommes as chairman of AT&T in the spring of 1972. DeButts was a very different sort of person than his predecessor—outgoing, aggressive, and with thirty years' experience primarily in the Bell operating companies (plus a year with AT&T government affairs in Washington).[120] Just a month after his appointment, deButts convened a presidents' conference to discuss the twenty-three unresolved issues still pending before the EPC. By the close of that conference, nothing had been decided, but deButts had made the "decision to decide." In Bell System lore, this meant that a clear strategic choice for dealing with the phenomenon of competition would finally be made at 195 Broadway. The bureaucracy that headed the nation's largest corporation was at last committed to making its move.[121]

Indications of deButts's direction were soon forthcoming. In the equipment segment of the market, for example, AT&T's critics had been urging substitution of a certification program for the cumbersome protective device; the FCC had proposed such a program and formulated a lengthy set of technical standards. In March 1973, however, the EPC recommended to the Bell operating companies a uniform policy of opposition to certification.[122] AT&T was as yet unwilling to cede CPE to unrestrained competition.

In the more contentious private line segment of the market, AT&T had been struggling to find ways of dealing with MCI in particular and the FCC's more general policy in *Specialized Common Carrier Services*. AT&T's first cut at the issue of competitive pricing was to lower rates only for those routes which specialized carriers planned to serve. Wisely, the EPC rejected this idea, concerned with the appearance of discrimination and the impracticality of chasing new entrants, route by route, with separate tariff filings. Indeed, it was disappointment with this decision, expressed at the 1972 president's conference, that emphasized for deButts the clear need to decide.[123]

For more than a year AT&T management had been considering an alternate plan—the Hi/Lo Tariff. In January 1973 the EPC recommended (and deButts approved) that Hi/Lo be filed with the FCC. This tariff proposed a break from national average pricing. AT&T designated certain cities as "high-density rate centers," others as "low-density rate centers." Rates for channels among the former would be $.85 per month per mile;

rates among the latter, $2.50. These rates, according to AT&T, were cost justified on a fully distributed, much less incremental basis.[124] Not surprisingly, the high-density centers were generally the markets in which MCI and other new entrants were active. The signal was clear: AT&T was getting tough.

In September 1973, John deButts took the occasion of the annual meeting of the National Association of Regulatory Utility Commissioners (NARUC) to announce his decision. His speech, reiterating a theme made forty-six years earlier by an AT&T chairman, was entitled "An Unusual Obligation." Addressing an audience of state and federal regulators, among whom was Bernard Strassburg, deButts attributed the onset of competition to "entrepreneurs who see opportunities for profit in serving selected segments"; to "manufacturers who seek to supplant the regulated common carriers in supplying the terminals for the common carrier network"; to some customers, especially "large businesses who see advantage to themselves in the new pricing arrangements" but "who have no obligation"; and finally to "some members of the regulatory community itself." Now the Bell System would take a stand:

> What do we believe?
> We believe that if competition is, as its advocates allege, the wave of the future in telecommunications, our responsibilities then require us to be as effective competitors as we know how to be. . . . At the same time we have recognized that the urgencies of competition do not exempt us from responsibility for rigorous examination of its long-term consequences. . . .
> We believe that the people of this country have been well served—and will continue to be best served—by a concept that has provided it the most highly developed communications service in the world, the concept of a universal system. . . .
> We believe that these services are so "affected with the public interest" as to require that their prices and the profits that derive from them be subject to continuing public regulation.
> And we believe, too, that the public interest . . . cannot help but be impaired by the duplication of facilities and the division of responsibility that will inevitably ensue from the further encroachment of competition. . . . In short, there is something right about the common carrier principle. There is something right about regulation. And—given the nature of our industry—there is something right about monopoly—regulated monopoly.
> The time has come to alert the public to what the public is largely unaware of—and that is that regulatory decisions have already been taken in its name that, whatever advantages they may afford for some people, cannot help but in the long run hurt most people.[125]

The Bell System, in other words, would neither concede the market to competition nor concede to the concept of competition.

In the marketplace, AT&T would continue to meet whatever competition was allowed in terminal equipment, switching equipment, and especially common carrier services. In Washington, AT&T called for a moratorium on regulatory experimentalism and a thorough assessment of the costs and benefits of competition versus monopoly.

This challenge to the FCC and the call for public debate only hardened the hearts of the Bell System's critics. The FCC did open a major docket to examine competition, but meanwhile Strassburg stepped up his support for MCI.[126] McGowen commenced a lobbying campaign at the Justice Department and in March 1974 filed a civil antitrust suit against AT&T. Initially, this action was an irritation but little more than that. DeButts felt secure in his company's conduct and its good relations with the Ford administration. It therefore came as a shock to almost everyone when Attorney General William Saxby inexplicably and unceremoniously filed an antitrust suit against AT&T on November 20, 1974.[127]

The antitrust challenge, no matter how remote, lent some urgency to deButts's call for legislative reform of the Communications Act. Although some AT&T executives thought it time to embrace competition, most probably agreed with deButts's line of defending regulated monopoly. Thus, legislation was drafted in cooperation with the independent phone companies and with the support of NARUC and the Communications Workers of America. It was introduced in 1976 as the Consumer Communications Reform Act. Despite its apparent broad base of support from nearly 200 cosponsors, the bill was universally dubbed the Bell Bill.

The Consumer Communications Reform Act was not written as a vehicle for discussion. It confirmed the existing rate structure, raised entry barriers in long distance, allowed Bell to adopt incremental pricing in competitive segments, and transferred responsibility for customer-provided terminal equipment from the FCC to state regulators—all in the name of protecting universal service.

The Bell Bill never made it past the House Subcommittee on Communications. The subcommittee chairman, Representative Lionel Van Deerlin (D–Calif.), had evidently not been cultivated. Although Van Deerlin did hold hearings in 1976, they were on the issue of telecommunications policy in general, not the bill per se. This move sidetracked the Bell Bill's early momentum, and in the next session Van Deerlin started over with a clean slate. In 1978, Van Deerlin and Representative Louis Frey (R–Fla.) introduced their own legislation to encourage competition, decrease regulation, and require vertically integrated exchange carriers (e.g., AT&T and GTE) to divest their manufacturing units. Over the next several years, Congress considered a series of initiatives to encourage competition. By 1981, when the Senate actually passed a telecommunications bill, S.898 (by a margin of ninety to four), it bore no resemblance to its genesis.[128]

Charles L. Brown succeeded John DeButts as chairman of AT&T in February 1979. His predecessor's strategy for getting a clear public policy decision on monopoly versus competition had failed. The Bell Bill was dead, the FCC's certification program for terminal equipment had been implemented, MCI's Execunet service was growing fast, and always in the background, *United States v. AT&T* ground steadily on. District Judge Harold H. Greene, who had recently assumed responsibility for the case from the ailing Judge Joseph Waddy, asserted exclusive jurisdiction, thereby shutting out the FCC.

By the time Brown took over, he was convinced that AT&T's clinging to monopoly while unable to compete was hopelessly draining the corporation of its energies, its spirit, and the public's respect. Brown came to the chairmanship from Illinois Bell and more recently as president of AT&T. His leadership style seemed to combine deButts's operating company perspective and commitment to action with Rommes's quieter, more reflective manner. Close observers invariably describe Brown as a "private person."[129]

Without fanfare or hesitation, Brown moved to take charge of the Bell System's course. "AT&T is ready without preconceptions," said Brown, "to explore alternative futures."[130] Even before assuming the chairmanship, Brown had established a planning council to rethink both strategy and organizational structure. He asked McKinsey & Company, the consulting firm that had earlier (in 1973) recommended a market-oriented reorganization, to restudy the situation. It did, and this time the consultants proposed an even more thorough restructuring, from a traditional functional organization, well suited to a service-oriented utility, to a product-line organization. A matrix

Chart 2-9
Evolution of Organization Structure at AT&T, 1907–1978

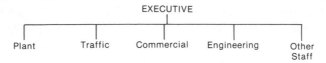

Traditional Functional Structure (1907–1973)

EXECUTIVE — Plant | Traffic | Commercial | Engineering | Other Staff

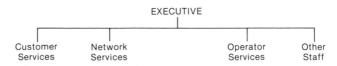

1973 Reorganization—System Structure

EXECUTIVE — Customer Services | Network Services | Operator Services | Other Staff

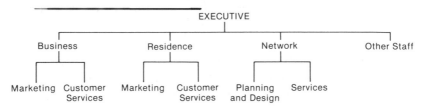

1978 Reorganization—Market Segment Structure

EXECUTIVE — Business (Marketing, Customer Services) | Residence (Marketing, Customer Services) | Network (Planning and Design, Services) | Other Staff

Source: AT&T: Adaptation in Progress (B) (Boston: HBS Case Services, 9-479-041), 1979, p. 5.

version of this restructuring was adopted in 1979, superimposing the product-line segmentation—business, residence, and network services—over the existing structure (see Chart 2-9).[131] But before the plan could even be implemented, the FCC ruled in *Computer Inquiry II,* thereby forcing yet another, and a more complicated, reorganization.

Externally, Brown made new and more constructive overtures to Congress, hoping to find a legislative compromise that might at once foreclose the antitrust suit and provide some feasible means of implementing the structural requirements of *Computer Inquiry II.* The Telecommunications Competition and Deregulation Act (S.898), although hardly ideal from AT&T's perspective, was the only game in town. Bell managers objected to its requirement that AT&T provide other common carriers with "equal access"—interconnection with local exchange networks on the same terms as AT&T Long Lines. They also disliked the severe new restrictions imposed on the regulated companies; the operating companies would, for example, have to buy a certain percentage of equipment from vendors other than Western Electric. And conversely, AT&T's new separate subsidiary (required by *Computer Inquiry II*) would not be allowed to benefit from any R&D that was funded by regulated activities.[132] Brown called S.898 "tough, double tough, on the Bell System because in mandating competition in the marketplace, it puts the reins on us, but gives spurs to everyone else."[133] Still, he accepted it.[134]

Legislative developments in the House, however, were still less sympathetic to AT&T. Representative Timothy Wirth (D–Colo.), who had succeeded Van Deerlin as chairman of the Communications Subcommittee, was highly critical of *Computer Inquiry II*, S.898, and the Bell System. Late in 1981, Wirth introduced his own bill, the Telecommunications Act of 1981, so different (and more restrictive on AT&T) from the Senate bill that immediate progress appeared impossible.

Brown's room to maneuver was now shrinking rapidly. In Federal District Court of the District of Columbia, Judge Harold Greene had surprised observers by tenaciously moving the antitrust case through discovery, stipulations, contentions, and proof to trial on January 15, 1981. One month later, William Baxter, a Stanford University law professor and conservative free market theorist, took charge of the Antitrust Division and vowed to litigate the AT&T case "to the eyeballs."[135]

The government's case was based on thousands of documents, hundreds of witnesses, and thirty specific acts that allegedly violated the antitrust laws. But the theory of the case, to which Baxter was intransigently committed, was that AT&T's vertical integration of regulated and nonregulated activities was inherently anticompetitive and conducive to predatory behavior: ownership of Western Electric had facilitated monopolization of equipment markets, just as ownership of local exchange companies ("the bottleneck facility") had cross-subsidized and facilitated monopolization of potentially competitive long-distance markets.[136] Philosophically, little had evidently changed since the Kingsbury Commitment or the 1949 antitrust case—only the government's resolve.

In September, Judge Greene ruled on AT&T's final request for dismissal. The judge denied it, commenting that "the Bell System has violated the antitrust laws in a number of ways over a lengthy period of time."[137] For those involved in the case, the handwriting was on the wall. In a speech a week later, Brown summed up the situation with painfully obvious frustration; between Congress and the courts, he said, "We are caught on the horns of a dilemma, that were it not so serious, would be ludicrous. It is time for somebody to act."[138]

Over the next two months, Brown made the momentous strategic decision to accept divestiture of the Bell System in exchange for what he believed was the right to compete freely in the information age. The settlement was announced by Brown and Baxter on January 8, 1982. AT&T would retain Long Lines, Western Electric, and Bell Laboratories but would divest the twenty-two telephone operating companies. It would also guarantee equal access for all competing long-distance carriers to the local Bell operating companies. The government, for its part, would rescind the restrictions in the 1956 consent decree.

Regulation and Competition

The divestiture of AT&T was the biggest, most complex restructuring in the history of business. The breakup took effect, after two years of preparations, on January 1, 1984. The creation of eight separate $16 billion companies dramatically interrupted the evolutionary course of deregulation; network engineering, market structure, competitive relationships, and political interests were thoroughly recast.

The broad terms of the consent decree were the product of tense, hasty negotiations between lawyers for AT&T and the Justice Department. The decree said almost nothing about asset assignment, corporate organization, network organization, or product-market restrictions. Those "details" were left to Bell System management, under the scrutiny of Judge Greene. There were several key elements to the outcome of this complex process. First, the twenty-two divested Bell operating companies (BOCs) were reorganized into seven regional holding companies (see Chart 2-10 and Table 2-6). The presidential study group that made this recommendation was charged with choosing "a structure which will optimize the ability of the Local Exchange Companies to provide high quality telephone service, earn well,

Chart 2-10

Regional Bell Operating Companies' Geographic Reorganization

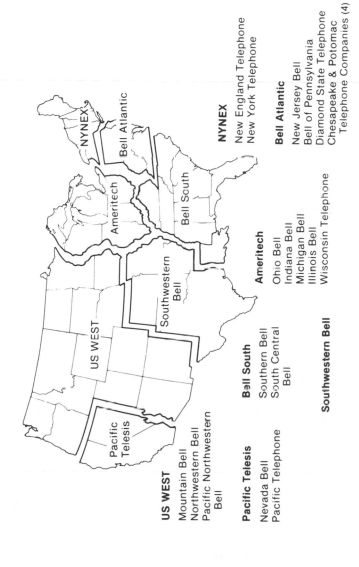

US WEST

Mountain Bell
Northwestern Bell
Pacific Northwestern
    Bell

Pacific Telesis

Nevada Bell
Pacific Telephone

Bell South

Southern Bell
South Central
    Bell

Southwestern Bell

Ameritech

Ohio Bell
Indiana Bell
Michigan Bell
Illinois Bell
Wisconsin Telephone

NYNEX

New England Telephone
New York Telephone

Bell Atlantic

New Jersey Bell
Bell of Pennsylvania
Diamond State Telephone
Chesapeake & Potomac
    Telephone Companies (4)

Source: Vietor and Dyer, *Telecommunications in Transition*, p. 15.

Table 2-6

Regional Holding Companies After Divestiture
Assets, Revenue, Employees, Leadership

Company	Assets	Revenues	Employees	Chf. Exec. Off.
	Dollars in Millions			
Ameritech	$16,257.0	$8,344.0	79,000	W. L. Weiss
Bell Atlantic	$16,264.1	$8,323.1	80,000	T. E. Bolger
BellSouth	$20,808.8	$9,799.1	99,100	W. R. Bunn
NYNEX	$17,389.0	$9,825.2	98,200	D. C. Staley
Pacific Telesis	$16,190.8	$8,082.1	82,000	D. E. Guinn
Southwestern Bell	$15,507.4	$7,754.9	74,700	Z. E. Barnes
US West	$15,053.6	$7,436.8	75,000	J. A. MacAllister

Source: W.B. Tunstall, *Disconnecting Parties* (New York: McGraw-Hill, 1985), p. 130.

and maintain access to capital markets."[139] Second, Bell Communications Research, or Bellcore, was organized as a joint venture to provide research, training, and supply services on a sufficient scale for the seven new regional companies. Third, exchange operations were reorganized into 161 Local Access and Transport Areas (LATAs), which were to become the basic units of local telephone service, each presumably defining a local monopoly market. By September 1, 1986, the operating companies were required to provide nondiscriminatory access, on an unbundled, tariffed basis, to firms competing in the market for interLATA service. And finally, the remaining units that would constitute the new AT&T—Long Lines, Bell Labs, and Western Electric—had to be re-formed into an organization that would comply with the terms of *Computer Inquiry II*. This new structure is represented in Chart 2-11 and Table 2-7.

In his Modified Final Judgment (MFJ), Judge Greene added several qualifications to the divestiture that have had a major impact on subsequent competition and regulation. Despite the rationale for divestiture (e.g., separating the naturally monopolistic bottleneck operations from naturally competitive ones), Judge Greene assigned the Yellow Pages, as an immensely profitable source of revenue, to the regional holding companies. He also required a significant portion of Bell Labs' human resources to be transferred to the new Bellcore. But of greatest significance were his restrictions on the future activities of the

Chart 2-11

AT&T Organizational Structure, Before and After Divestiture

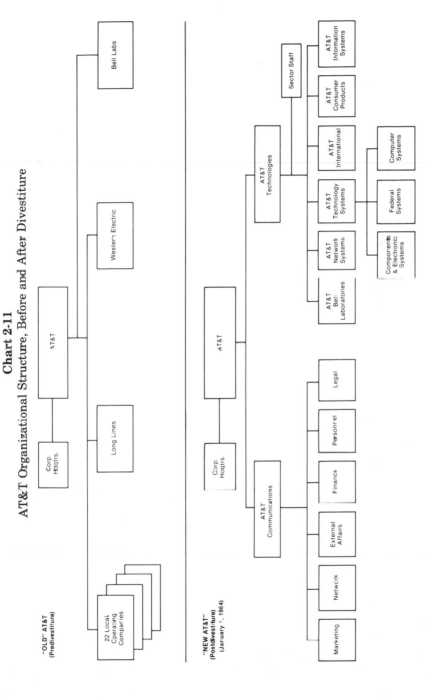

Source: W.B. Tunstall, *Disconnecting Parties* (New York: McGraw-Hill, 1985), p. 118.

Table 2-7
AT&T Financial Structure, Before and After Divestiture

	Old AT&T 1983[a]	*New AT&T* 1984[b]
Total operating revenues	$ 67,599	$ 56,544[c]
Total operating expenses	47,160	51,565
Net operating revenues	20,439	4,976
Net income	7,188	2,110
Annual dividend (per share)	$5.40	$1.20
Per share earnings	$7.88	$2.02
Number of shares (weighted average)	894.1	989.1
Total assets	$153,510	$ 34,277
Shareholders' equity	65,147	13,229
Long-term and intermediate debt	45,320	9,469
Current liabilities	40,700	6,764
Employees	992,000	385,000

a. For the twelve-month period ended June 30, 1983 (unaudited).
b. Revenue, income, employee, and dividend data are estimates for 1984; assets and liabilities are consolidated on a pro forma basis from June 30, 1983, to reflect divestiture.
c. Note that this financial forecast is for newly restructured entities with new modes of operation and new tariff structures for which comparable historical results do not exist. The extent of the changes required as a result of divestiture do not permit meaningful comparisons between forecasted results and the results for periods prior to 1984. In addition, the 1984 financial forecast reflects significant changes in accounting policies, including the consolidation of Western Electric and the adoption of accounting policies appropriate for nonregulated enterprises. As an example, "total operating revenues" include revenues collected to pay NTS subsidies of local exchange facilities.
Source: AT&T Information Statement and Prospectus, November 8, 1983. Cited in R. Vietor and D. Dyer, *Telecommunications in Transition* (Boston: HBS Case Services, 1986), p. 101.

BOCs. In the absence of a waiver from the court, no BOC or any affiliated enterprise could:

1. Provide interexchange telecommunications services or information services;
2. Manufacture or provide telecommunications products or customer premises equipment; or
3. Provide any other product or service, except exchange telecommunications and exchange access service, that is not a natural monopoly service actually regulated by tariff.[140]

In the aftermath of this divestiture, competitive forces and regulatory constraints clashed so fundamentally that even the most carefully construed transitional arrangements would prove untenable. Five sets of issues have since dominated the arena of telecommunications policy: (a) access charges (jurisdictional cost separations), (b) state price and service regulations,

(c) enhanced services, (d) asymmetrical regulation of AT&T (but not the other long-distance carriers), and (e) the MFJ's restrictions. A sixth issue, that of asymmetrical international competition in the markets for telecommunications equipment and enhanced services, looms larger every month.

Access Charges

Even before divestiture, the system of jurisdictional separations and division of revenues that had subsidized universal service was coming apart. Competition in the interLATA (long-distance) markets had already necessitated "rough justice," in the form of ENFIA. And *bypass* of the local exchange by large users had begun to erode the revenue base of the public switched network. By requiring that access to the local exchange be nondiscriminatory and unbundled, the MFJ forced the FCC to devise a new system of pricing and cost recovery.

Shortly after approving the ENFIA tariff in 1978, the FCC had opened a wide-ranging investigation of the entire long-distance (MTS and WATS) market structure.[141] The discount arrangement for the access fees paid by other (i.e., non-Bell) common carriers (OCCs) was at best temporary. AT&T, despite its advantage of superior quality access and its immense market power, was competing at a significant cost/price disadvantage. Over the years, as Chart 2-4 indicated, the proportion of non-traffic-sensitive exchange plant costs allocated to the interstate jurisdiction had grown to 26 percent. This was 3.3 times the proportion of long-distance to local usage and had created a system of prices that did not in any way reflect costs. In effect, one-fourth of the fixed costs of the local exchange networks was recovered from the usage-sensitive toll charges for long-distance service.

To the extent that long-distance usage and local exchange plant costs were attributable to different customers with different demand elasticities, this system entailed immense subsidies and distortions of the patterns of supply and demand. In general terms, business customers, especially the large corporations that accounted for the lion's share of long-distance usage, were subsidizing residential service; urban customers were subsidizing rural service; the East and West coasts were subsidizing the Midwest; and anyone who made more long-distance calls than average was subsidizing anyone who made fewer than average. This last group was *thought* to include low-income users, minorities, the elderly, and the poor.[142]

By 1984, AT&T's access costs (both interstate and intrastate) were $15.5 billion, about 63 percent of its long-distance operating costs; this translated into 15 out of every 30 cents that AT&T received per average minute of use. By contrast, MCI and the other new entrants were paying about 5 cents per minute for access.

The FCC's search for an alternate cost-recovery and access pricing mechanism lasted five years. The proceedings encompassed the views of virtually every interested party. No option was overlooked. At one extreme, all fixed costs of the local networks could be shifted to end users, on a flat rate (per month) basis. At the other extreme, they could continue to be paid by long-distance carriers entirely on a usage basis. Or there could be some mix of these two cost-recovery methods, in various proportions, phased in over time.[143] The pace of the investigation quickened once the MFJ was issued, to comply with Judge Greene's deadline for equal access.

In December 1982, the commission announced its Access Charge Plan, citing four objectives: (a) greater economic efficiency, (b) prevention of *uneconomic* bypass (bypass that exceeded AT&T's costs but not its price), (c) a less discriminatory pricing system, suitable to competition, and (d) preservation of universal service. The plan had three parts (see Chart 2-12):

1. *Customer access line charge:* Every local subscriber would pay a fixed monthly fee per line (initially $2 for residential customers, and $6 for business customers), increasing over several years to $6 or $8.
2. *Common carrier line charge:* Long-distance carriers would continue to pay a variable, per-minute-of-use access fee, which would gradually be phased out, as the customer charge was phased in, until 1990, when they would pay only for traffic-sensitive costs. The discount provided to OCCs under ENFIA would be reduced to 35 percent initially and then phased out gradually over thirty months as they received equal access.
3. *Universal service fund:* To assist local phone companies with unusually high fixed costs, the plan would create a

Chart 2-12
Original FCC Access Charge Plan, 1983

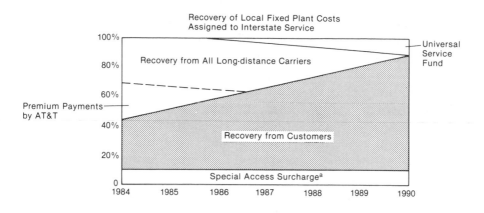

a. On special long-distance service.
Source: AT&T Communications.

fund to which long-distance carriers contributed (on a usage basis).[144]

By realigning costs, forcing "cost causers" to pay, and reducing the competitive handicap on which MCI and the other new long-distance carriers depended, the FCC's Access Charge Plan sparked a major political battle that lasted throughout 1983. To some congressmen, it appeared that the FCC was preempting the policymaking prerogatives of Congress. For this reason among others, Representatives Timothy Wirth and John Dingell (chairman of the House Commerce Committee) and Senator Robert Packwood (chairman of the Senate Commerce Committee) tried to block the plan, on a jurisdictional basis if nothing else. State regulators, moreover, felt that the FCC was interfering in local pricing (e.g., a $2 per month charge for local access), and so they, too, opposed it on jurisdictional as well as substantive grounds.[145]

In hearings conducted by both houses of Congress, consumer activists, state utility regulators, organized labor, small businesses, and the other common carriers bitterly contested the plan.[146] The Reagan administration laid low, leaving only large corporate users, the FCC, AT&T, and the emerging Bell companies to support the plan. And since AT&T and the Bell companies were preoccupied at the time with immense problems of implementing divestiture, their political effectiveness was minimal.

This fight came to a head in October 1983, when AT&T and the seven new Bell companies filed tariffs for access under the new system. The plan's opponents claimed that proposed reductions in long-distance rates did not fully reflect AT&T's reduction in (access) costs—the difference amounting to a "Great Phone Robbery."[147] MCI and the other new competitors added to the pressure with a letter to the FCC chairman, Mark Fowler. In this so-called "Eight-Carrier Letter" they claimed that any immediate reduction in their discount, and its phasing out over thirty months, would be financially ruinous. Their collective profits would plunge from an estimated $484 million (for 1984) to losses of more than $500 million.[148] Together, these allegations gave political momentum to Wirth's legislation, which passed the House by a voice vote. This bill, the proposed Universal Telephone Service Preservation Act, would have prevented the FCC from shifting fixed costs back to end users. It would also have established mandatory "lifeline" rates (minimum bills) for the poor and elderly and would have required large users that bypassed the network to contribute to local telephone service anyway.[149]

When thirty senators cosigned a letter from Robert Dole to the FCC, urging reconsideration and delay (until after elections), Fowler knew his plan for access charges was politically dead, at least in its original form. He immediately postponed the customer charge until mid-1985, agreed to a cap of $4, and restored the access discount for AT&T's competitors to 55 percent (see Chart 2-13).[150]

Between 1985 and 1987, the commission made some limited progress with its efforts to reform access cost recovery. A business-line charge ($6 per month) did take effect on schedule, and in mid-1985 a charge of $1 per month was imposed on residential customers. This charge was raised to $2 in 1986 and is currently scheduled to rise slowly to $3.50 by 1989. As a result of these charges, long-distance rates have been reduced more than 30 percent to date and should fall by another 10 to 15 percent.[151]

Restoration of the discount for MCI, Sprint, and the other new

Chart 2-13
Recovery of Local Fixed Plant Costs Assigned to Interstate Service

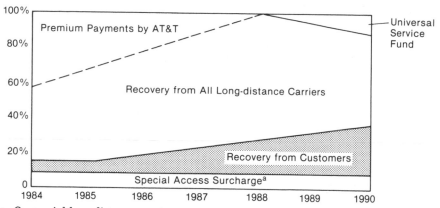

a. On special long-distance service.
Source: AT&T Communications.

long-distance competitors turned out to have a minimal effect. In the "equal access" competitions, where subscribers were required (exchange by exchange, as equal access connections were installed) to choose by ballot their preferred long-distance carriers, AT&T was surprisingly successful. Customer loyalty and indifference helped, but so did an extraordinary marketing campaign, headed by actor Cliff Robertson's image advertising. As of 1987, AT&T retained about 84 percent of interexchange traffic; MCI had 8 percent, Sprint 4 percent, and all others 4 percent.[152]

State Regulation

Divestiture, combined with federal regulatory changes, created innumerable problems for state regulators. In 1983 and 1984, public service commissions everywhere had to arrange new reporting systems for the regional holding companies and conduct rate cases for the rate hikes requested by the operating companies to reflect their accelerated depreciation schedules. Before these issues were resolved, state regulators had to begin considering requests for rate deaveraging and measured service; intrastate access charges; new service offerings; bypass; and competition in the markets for intrastate, interLATA toll service, intraLATA toll service, and even local exchange services.

By July 1986, according to a survey conducted by the National

Telecommunications and Information Administration (NTIA), regulatory change was well under way but diverged widely in substance and process from state to state. In Virginia, for example, interLATA competition was not only allowed but deregulated in 1984. In Utah, on the other hand, competition in interLATA toll was explicitly prohibited. The Nebraska legislature enacted a law to deregulate intraLATA service (terminating rate-of-return regulation) but prohibited competition until 1989. Louise McCarren, chairman of the Vermont Public Service Board, proposed a "social contract" whereby New England Telephone would commit to residential rate stability in return for competitive flexibility in business service markets. In Illinois, the Commerce Commission unbundled access from intrastate toll service, adopting a flat rate, subscriber-line charge capped at $5.52. Idaho, by contrast, explicitly prohibited such a charge.

In the aggregate, thirty-six states (out of thirty-eight with more than one LATA) had permitted interLATA competition; fourteen states had permitted intraLATA toll competition. Pricing flexibility in toll services was introduced in twenty-eight states; pricing for local exchange services, in thirty-five states. Six states had implemented access (subscriber-line) charges, and fifteen states had approved limited forms of detariffing "competitive" services.[153]

Enhanced Services

The FCC had scarcely ruled in *Computer Inquiry II* when divestiture released AT&T from the terms of the 1956 consent decree. And anyway, neither the Justice Department nor AT&T had been satisfied with the FCC's structural solution, albeit for entirely different reasons; if separate subsidiaries could effectively prevent cross-subsidies between regulated and competitive markets and if they were workable from an operational viewpoint, why had divestiture been necessary?

With eight huge companies aggressively digitizing their networks, making possible an array of new services and equipment, the distinction between competitive and noncompetitive assets and operations was again blurring faster than the FCC could implement its rules. Since AT&T had divested its bottleneck facilities (i.e., the local exchange operating companies), it sought relief from the strictures of *Computer Inquiry II*.[154] The

FCC undertook a Third Computer Inquiry and announced its decision in June 1986. The commission concluded that with by-pass technologies proliferating and competition increasingly robust, the threat of cross-subsidy was minimal. It ruled that structural safeguards could be replaced by "nonstructural" safeguards, of which the most important was the implementation of "open network architecture." Access to local and interexchange networks would be fully unbundled from competitive services and products by the provider and would be made available on equal technical and pricing terms to any potential user (competitor).[155]

Asymmetrical Regulation

Even after divestiture, the new AT&T-Communications (renamed from Long Lines) remained fully regulated by the FCC as to rates, rate of return, services, and facilities. Of course, it still had tremendous market power in the interexchange business and preferential (high-quality) access to most exchanges until late 1986. Still, none of the other long-distance carriers with which AT&T competed, such as MCI and Sprint, were regulated as common carriers.[156] They could enter and exit markets freely and selectively, build facilities and change prices without the FCC's prior approval, and offer new services on short notice.

Until 1986, the other long-distance companies remained adamantly opposed to even the most gradual regulatory relief for AT&T.[157] They claimed that AT&T had deep pockets for advertising, a huge reserve of goodwill with customers, bargaining power with the operating companies, and overwhelming market presence. MCI suggested that deregulation would be appropriate sometime in the mid-1990s, after AT&T's market share had fallen to 40 percent.[158] Sprint went so far as to admit that AT&T's unit costs, without AT&T's access cost disadvantage, were 20 to 30 percent below its own. According to Sprint, were AT&T deregulated, its competitors would never earn their cost of capital and would more than likely be driven from the market.[159]

By mid-1986, however, equal access had been implemented and the other carriers' cost advantage over AT&T (from their access cost discount) was largely dissipated. Since the FCC would not allow AT&T, as a public utility, to raise its rate of return (above 12.2 percent), AT&T was forced to keep lowering rates as customer access charges were implemented. This put immense price pressure on MCI and the others, destroying their own real competitive advantage. Early in 1987, MCI announced losses of $448 million, and in an irony of epic proportions, called for the immediate deregulation of AT&T. Bert Roberts, Jr., MCI's president, urged the FCC to "redirect its resources to regulate the true monopolies in this industry—the local telephone companies."[160]

But there was a double irony here. As AT&T's access cost handicap had dissipated, revealing its immense economies of scale and scope, competition in the interexchange market began to appear untenable—a failed experiment—*unless* the huge Bell operating companies were freed to enter the interLATA market.

The MFJ Restrictions

At the time of the divestiture, Judge Harold Greene had ordered that the Justice Department review the MFJ every three years to ensure its continuing relevance to the state of technology and competition. In the meantime, from all of the regional Bell holding companies there had been a constant clamor for waivers in order to enter unregulated lines of business: international operations, cellular mobile phones, electronic publishing, computer retailing, real estate, and financial services were just a few.

The Justice Department commissioned Dr. Peter Huber, an engineer-lawyer who had left the faculty at Massachusetts Institute of Technology, to study the telecommunications industry and recommend changes, if appropriate, in the MFJ. In January 1987, Huber submitted his report, recommending that the Bell operating companies be freed from restrictions on manufacturing equipment and providing enhanced services. He based the recommendations, in part, on the striking view that the emerging architecture of the network (discussed earlier in this chapter) warranted (oligopolistic) competition by vertically integrated firms.[161] The report also documented competitive conditions, segment by segment throughout the telecommunications industry, in access lines, fiber optics, telephones, PBXs, local area networks, mobile services, packet switching, information services, electronic mail, and so on.

The Justice Department took Huber's advice and then some, apparently rejecting its previous stance and a part of the theory of its case for AT&T's divestiture. In February 1987, it recommended abandonment of the MFJ restrictions on equipment manufacturing, information services, *and* entry into interLATA long-distance markets, except for service originating or terminating in the operating company's own local exchange territory.[162]

It remained to be seen what Judge Greene would do.

Regulation and Deregulation in Perspective

For more than seventy years, the market structure of telecommunications has been defined by the interaction of technology, strategy, and public policy. Above all, this has been a dynamic relationship. Technology, from the earliest years of telephony, was the source of competitive advantage. The technical characteristics of a rudimentary network set the Bell System on a course of horizontal integration, vertical integration, and end-to-end voice service. Strategy, as early Bell managers used it, was a means of sustaining their competitive advantage beyond their first-mover advantage in the face of easy entry and intense competition. But a strategy that was so successful yet denied interconnect in pursuit of consolidation soon ran afoul of public policy. At the height of the Progressive Era and again during the Great Depression, Bell's exercise of market power was unacceptable to the Justice Department, to state legislatures, and eventually to Congress. Vail was the first of several Bell chief executives who chose regulation and monopoly over competition as the best means for organizing the network and controlling the growth of the telecommunications business.

In this chapter I have tried to understand the process of regulation, from inception to partial remission, as *a series of institutional arrangements designed politically to "structure" telecommunications markets*.[163] Almost by definition, the objectives of regulation clashed with the economic characteristics of technology and with the successes of business strategy.

The starting point in this "structuring process" is the basic economic and political conditions that form the context for some sort of microeconomic intervention by government. The Progressive Era provided such a context for the Kingsbury Commitment; the Great Depression was the context for the Communications Act; and the post-Vietnam stagflation and consumer activism set the stage for deregulation.

Of course, the occasion for regulatory change is partially industry specific. In telecommunications, the perception of market imperfections that derived from the state of technology *at the time* gave rise to commission regulation. Positive externalities (the network as a public good) and natural monopoly were the implicit justifications for state and federal regulation. Meanwhile, the way in which AT&T managed those characteristics repeatedly gave rise to antitrust enforcement.

From the 1920s to the mid-1950s, the implementation of government regulation increasingly shaped the structure of telecommunications markets. This structuring process began with the Justice Department's definition of the boundaries of the markets, continued with state regulatory and judicial decisions on value-of-service pricing, and went on to the FCC's policies for encouraging universal, high-quality service through national average pricing and the subsidy from toll usage to the fixed costs of local plant.

The market structure created by this regulatory process had a political analogue of interests vested in the status quo: state regulators, AT&T and its operating units, the Communications Workers of America, AT&T stockholders, hundreds of interconnecting independent phone companies, and the huge mass of relatively high-cost, relatively low-usage residential customers that were subsidized by large corporate customers.

The process of change and competitive ventilation that began after 1956 and accelerated in the late 1960s was driven by regulatory failures and entrepreneurship, changes in basic economic and political conditions, and changes in market structure that stemmed from technological innovations in transmission and electronic switching. With new technologies breaking down regulatory entry barriers and market segmentation, and given the elaborate system of cross-subsidies that prevailed, entrepreneurs seized the opportunities for entry and innovation. Efforts by regulators to accommodate and control these developments not only failed but also hastened the process of deregulation.

The fact that AT&T's competitive responses were repeatedly challenged as illegal only underscores the immense imbalance that had developed between economic and institutional realities.

Technological innovation in electronics, together with the widening gulf between regulated prices and economic costs, made significant regulatory change and market restructuring occur. But the course it has taken and the effectiveness of the process were a function of institutions and individuals. And in my view, neither the process nor the results have been very successful to date.

The best results, I think, have been in the terminal equipment sector, where competition and detariffing have stimulated diversity, innovation, and a whole array of new service possibilities. But even here, the alarming rate of import penetration, from 16 to 24 percent since 1980 (and with no reciprocal export growth), poses grim prospects for the international competitiveness of American firms in this critical manufacturing sector.[164]

Yet the real source of difficulties for regulatory reform has been the huge gap between costs and cost causation that value-of-service pricing and jurisdictional separations created *and* the inability or unwillingness of deregulators, trustbusters, and managers to deal with it. The FCC, for example, embarked in the mid-1960s on a course of experimental entry and competition in the interexchange segment of the business. Yet neither the Common Carrier Bureau nor the majority of the commission would squarely address the policy implications of selective entry, the necessity for AT&T to deaverage rates, or its own obligation to develop an appropriate method of rate regulation. In this and in its commitment to gradual and limited entry, the commission evidently mistook its own artificial segmentation of the market by tariffs (e.g., MTS, WATS, PL [private line]) for real economic distinctions.

Likewise, the separations subsidy (and value-of-service pricing in general) contradicted Baxter's theory of the case. Baxter believed that bottleneck facilities were used unfairly to hamper entry into potentially competitive markets, holding rates above efficient levels. But of course it was those very services which cross-subsidized the local exchange bottleneck. Not surprisingly, Baxter's solution of separating local exchange from interexchange facilities by divestiture not only created pricing and distributive problems but also did not alleviate the competitive ones.

For AT&T management generally and for Brown in particular, the contradictions between value-of-service pricing and competitiveness posed intractable strategic problems. In the end, Brown acceded to divestiture but still failed in the short term to attain the status of unregulation necessary to compete freely. Residual market power and the continuing burden of tariffed access costs would obscure Brown's strategic vision on into the 1990s.

Finally, the pricing structure created by past policies continues to foil the market structuring dictum of Judge Harold Greene. The monopolistic segregation of the BOCs has been under constant pressure from continuing technological advances, from market pressures for integrated enhanced services, and from problems with related deregulation to date (such as the shaky competition in long-distance markets and the overwhelming foreign competition in equipment markets). The MFJ's limits on the BOCs, like previous restrictions on product market and entry, make no intrinsic sense in so dynamic an industry.

The mixed regime of regulation and competition that now prevails is scarcely more tenable, economically or politically, than it was in 1910 when Vail first addressed the issue. Regulation, as he explained then, means public discussion, deliberation, and nondiscrimination. These things are not just incompatible with but are antithetical to aggressive competition. In the transition of Vail's era, private and public policy made their choices for regulation and monopoly. Today, AT&T and the regional holding companies have generally chosen competition. Government must drastically reform cost-based rate regulation and entry restrictions or else deregulate.

Notes

1. Ernest Braun and Stuart MacDonald, *Revolution in Miniature: The History and Impact of Semiconductor Electronics* (Cambridge: Cambridge University Press, 1982), p. 33; and Jeremy Bernstein, *Three Degrees Above Zero* (New York: Mentor, 1984), pp. 82–114.

2. For the best scholarly analysis of AT&T's response to deregulation, see Peter Temin, *The Fall of the Bell System* (New York: Cambridge University Press, 1987); for a reliable popular version of this story, see Steve Coll, *The Deal of the Century* (New York: Atheneum, 1986); and see Charles Brown, "A

Personal Introduction," in *Disconnecting Bell: The Impact of the AT&T Divestiture*, ed. Harry M. Shooshan III (New York: Pergamon Press, 1984), pp. 1–7.

3. AT&T, *Annual Report, 1910*, pp. 22–23.

4. Neil Wasserman, *From Invention to Innovation: Long-Distance Telephone Transmission at the Turn of the Century* (Baltimore: Johns Hopkins University Press, 1985).

5. Drawing on comparisons with Sweden and France, Gerald Brock argues that Bell's monopoly position in the United States substantially slowed the industry's growth and that after competition commenced in 1894, real growth and market penetration took off; see Gerald Brock, *The Telecommunications Industry: The Dynamics of Market Structure* (Cambridge, Mass.: Harvard University Press, 1981), pp. 99–125, 143–147.

6. Robert Garnet, *The Telephone Enterprise: The Evolution of the Bell System's Horizontal Structure, 1876–1909* (Baltimore: Johns Hopkins University Press, 1985).

7. George Smith, *The Anatomy of a Business Strategy: Bell Western Electric, and the Origins of the American Telephone Industry* (Baltimore: Johns Hopkins University Press, 1985).

8. Quoted in Garnet, *The Telephone Enterprise*, p. 77.

9. *United States v. AT&T*, Civil Action No. 74-1698, "Defendants' Third Statement of Contentions and Proof," March 10, 1980, p. 139.

10. In the government's recent antitrust suit, AT&T argued that MCI and other competitive entrants in the long-distance market merely cream-skimmed the highest-volume, lowest-cost segments. A frequent rebuttal was that for decades AT&T itself had avoided the low-volume, least-lucrative regions, leaving them for independents and the Rural Electrification Administration to develop.

11. Harry B. MacMeal, *The Story of Independent Telephony* (Chicago: Independent Pioneer Telephone Association, 1934).

12. In 1897, for example, the National Independent Telephone Association was organized for mutual protection and development and for the promotion of independent long-distance connections. Several corporations were also organized, including the Telephone, Telegraph & Cable Company in 1899. This venture, plagued by continuous financial difficulties, was reorganized, bought out by a syndicate of Bell companies, and dissolved in 1902; Federal Communications Commission, *Investigations of the Telephone Industry in the United States*, in U.S. Congress, House, *House Document No. 340* (76th Cong., 1st Sess.), 1938, pp. 131–132.

13. FCC, *Telephone Investigation*, docket no. 1, vol. 65, "Report on Control of Independent Telephone Companies," June 1937, pp. 13–14.

14. FCC, *Telephone Investigation*, docket no. 1, vol. 66, "Financial Control of the Telephone Industry," June 1937, Chapter 4.

15. AT&T, *Annual Report, 1910*, p. 21.

16. *Congressional Record* 45 (1910): 5536.

17. Michigan Public Utilities Commission, *Citizens Telephone Co. of Grand Rapids*, P.U.R. 1921E, pp. 308, 315.

18. FCC, docket no. 1, "Control of Independents," p. 15.

19. AT&T, *Annual Report, 1910*, p. 36.

20. Defendants' Third Statement, p. 164. Vail and Kingsbury were named codefendants.

21. N.C. Kingsbury to J.C. McReynolds, J.C. McReynolds to N.C. Kingsbury, and W. Wilson to N.C. Kingsbury, December 19, 1913, in FCC, docket no. 1, vol. 65, pp. 34–40.

22. FCC, docket no. 1, "Control of Independents," pp. 41–42.

23. Defendants' Third Statement, p. 167.

24. *Congressional Record* 61 (1921): 1983.

25. U.S. Congress, Senate, Committee on Interstate Commerce, *Senate Report No. 75* (67th Cong., 1st Sess.), 1921, p. 1.

26. E.K. Hall, memorandum, June 14, 1922, in FCC, docket no. 1, vol. 65, pp. 50–51. This did not settle matters entirely, however. By 1925, independents were again dismayed by AT&T's aggressive merger strategy. Their association tried to obtain Bell's consent to a policy of relinquishments equal to acquisitions. AT&T declined, threatening to stop acquisitions altogether. For those independents which hoped to be acquired, this cure was worse than the disease.

27. Ibid., pp. 72–107.

28. Defendants' Third Statement, pp. 225–235.

29. *Historical Statistics of the United States,* series R 1–12.

30. FCC, *Investigations of the Telephone Industry,* pp. 39, 41; and, N.R. Danielian, *AT&T: The Story of Industrial Conquest* (New York: Vanguard Press, 1939), p. 410.

31. See Ellis Hawley, *The New Deal and the Problem of Monopoly* (Princeton: Princeton University Press, 1966), Chapters 12, 15–17.

32. FCC, *Investigations of the Telephone Industry,* p. 145.

33. *United States v. AT&T,* Civil Action No. 74-1698, "Plaintiff's Third Statement of Contention and Proof," January 10, 1980, p. 1831.

34. U.S. Congress, House, *House Report No. 1273* (73rd Cong., 2nd Sess.), "Report on Communications Companies" (Splawn report), part 3, no. 1, 1934, pp. ix, xii.

35. U.S. Congress, House, *Study of Communications by an Interdepartmental Committee* (73rd Cong., 2nd Sess.), Committee Print (Roper report) 1934, pp. 14, 24–25.

36. Although the need for centralized federal regulation was noncontroversial, there was nonetheless considerable disagreement over the new commission's specific powers. In the hearings, Walter Gifford, the president of AT&T, objected that several proposed extensions would interfere with needed managerial discretion; see U.S. Congress, House, Committee on Interstate and Foreign Commerce, *Hearings on H.R. 8301* (73rd Cong., 1st Sess.), 1934, pp. 165–206.

37. Bernard Schwartz, ed., *The Economic Regulation of Business and Industry: A Legislative History of U.S. Regulatory Agencies* (New York: Chelsea House, 1973), vol. 4, p. 2374.

38. Franklin D. Roosevelt, "Message to Congress, February 26, 1934," in U.S. Congress, House, *House Report No. 1850* (73rd Cong., 2nd Sess.), 1934, pp. 1–2.

39. Senator Dill, the bill's author, *Congressional Record* 78 (1934): 4139; and Representative Mayes, *Congressional Record* 78 (1934): 10990; quoted in G. Hamilton Loeb, *The Communications Act Policy Toward Competition: A Failure to Communicate* (Cambridge, Mass.: Center for Information Policy Research, 1977), p. 41.

40. Communications Act of 1934, Title II, in Schwartz, *Economic Regulation,* vol. 4, pp. 2383–2395.

41. Ibid., p. 2377.

42. Loeb, *The Communications Act,* pp. 51–92. The author provides a thorough explanation of how this controversy took shape during the 1970s and how its various dimensions related to the intent and interpretation of the act.

43. The rights and obligations of common carriage were articulated as early as 1768, by Blackstone (3 *W. Blackstone, Commentaries* 219): "If a ferry is erected on a river, so near another ancient ferry as to draw away its custom, it is a nuisance to the owner of the old one. For where there is a ferry by prescription, the owner is bound to keep it always in repair and readiness for the ease of all the king's subjects; otherwise he may be grievously amerced."

44. U.S. Congress, Senate, *Senate Report No. 781* (73rd Cong., 2nd Sess.), 1934, p. 2.

45. U.S. Congress, House, *Hearings on H.R. 8301*, p. 200.

46. FCC, *Telephone Investigation*, docket no. 1. Baker Library, at the Harvard Business School, Boston, Mass., has one of the few remaining sets of the original planographed staff studies, bound in seventy-seven volumes (YCG, U585), of which twelve appear to be missing. The library also has nine volumes of companion studies by the Telephone Rate and Investigation Department (YCG, Y585b).

47. FCC, docket no. 1, "Final Report of the Telephone Rate and Investigation Department," June 16, 1938, pp. 68–69.

48. Ibid., p. 72.

49. FCC, docket no. 1, "Control of Independents," p. 92.

50. Ibid., p. 119. By 1935, the long-distance market produced $240 million in gross revenues, of which $56 million was interchanged with independents. Of this, Bell kept 79 percent, returning only $12 million to independents; p. 110.

51. James W. Sichter, *Separations Procedures in the Telephone Industry: The Historical Origins of a Public Policy* (Cambridge, Mass.: Center for Information Policy Research, 1981), pp. 35–43.

52. *Smith v. Illinois Bell Telephone Co.*, 288 U.S. 133 (1930).

53. Sichter, *Separations Proceedings*, p. 72.

54. National Association of Railroad and Utility Commissioners, *Proceedings, 1947* (Boston), p. 355; and, Richard Gabel, *Development of Separations Principles in the Telephone Industry* (East Lansing: Institute of Public Utilities, Michigan State University, 1967), pp. 30–31.

55. National Association of Railroad and Utility Commissioners, *Proceedings, 1950* (Phoenix), p. 47.

56. FCC, *Annual Report, 1950*, pp. 46, 56.

57. Ibid., p. 57.

58. Quoted in Peter Temin and Geoffrey Peters, "Is History Stranger than Theory? The Origin of Telephone Separations," *American Economic Review* 75, no. 2 (May 1985): 324–327; and, Gabel, *Development of Separations Principles*, pp. 68–69.

59. National Association of Railroad and Utility Commissioners, *Proceedings, 1951* (Charleston), pp. 156, 166–167; and, FCC, *Annual Report, 1951*, pp. 47–48.

60. Defendants' Third Statement, p. 237.

61. FCC, *Annual Report, 1952*, pp. 43–44.

62. FCC, docket no. 1, "Control of Independents," pp. 63–64. Sixty percent of the operating companies' rate base—or 30 cents on each dollar collected from telephone subscribers—consisted of items furnished by Western Electric.

63. NARUC, *Proceedings, 1950*, p. 43.

64. Christopher Sterling et al., eds., *Decision to Divest: Major Documents in United States v. AT&T, 1974–1984* (Washington, D.C.: Communications Press, 1986), pp. i–8.

65. *United States v. Western Electric Company*, Civil Action No. 17–49, pp. 37–65, in Sterling, *Decision to Divest*, vol. 1, pp. 85–113.

66. Ibid., pp. 124–141.

67. Peter W. Huber, *The Geodesic Network* (1987 report on competition in the telephone industry), prepared for the Department of Justice in connection with the court's decision in *United States v. Western Electric* 552 F. Supp. 131 (D.D.C. 1982) (Washington, D.C.: Government Printing Office, 1987), Chapter 1.

68. Braun and MacDonald, *Revolution in Miniature*, pp. 196–197.

69. AT&T, *Events in Telecommunication History*, p. xi.

70. Huber, *Geodesic Network*, p. 1.18.

71. *In the Matter of Allocation of Frequencies in the Bands Above 890 Mc.* 27 FCC 359, 364 (1959).

72. Clare McGillem and William McLauchlan, *Hermes Bound: The Policy and Technology of Telecommunications* (West Lafayette, Ind.: Purdue University, 1978), pp. 42–43.

73. *United States v. AT&T*, C.A. No. 74-1698, Testimony of John Nemecek (Defendant's Exhibit D-T-366), pp. 5–6.

74. See, for example, Gale D. Merseth and Thomas McCraw, *Tele/Resources, Inc.*, 4-378-113. Boston: Harvard Business School, 1977.

75. *United States v. AT&T*, C.A. No. 74-1698, Testimony of Ronald Ruebusch (Defendant's Exhibit D-T-359), p. 6.

76. Huber, *Geodesic Network*, p. 1.3.

77. Defendants' Third Statement, pp. 1126–1130.

78. *Hush-a-Phone Corp.*, 20 F.C.C. 391, 424 (1955).

79. *Hush-a-Phone Corp. v. United States*, 238 F. 2d 266 (D.C. Cir. 1956).

80. Plaintiff's Third Statement, pp. 1096–1098.

81. *Carter v. American Telephone & Telegraph*, 250 F. Supp. 188 (N.D. Tex. 1966), aff'd, 365 F.2d 486 (5th Cir. 1966).

82. *Carterfone*, 13 F.C.C. 2d 606 (1968).

83. Plaintiff's Third Statement, pp. 1100–1101.

84. *"Foreign Attachments" Tariff Revisions*, 15 F.C.C. 2d 605 (1968).

85. 35 F.C.C. 2d 539 (1972); 56 F.C.C. 2d 593 (1975).

86. *Second Computer Inquiry*, 77 F.C.C. 2d 384, 390 (1980).

87. *Regulatory and Policy Problems Presented by the Interdependence of Computer and Communications Services and Facilities*, 28 F.C.C. 2d 267 (1971).

88. 77 F.C.C. 2d 384, 391, 430.

89. Ibid., 386–387, 438.

90. *Above 890 Mc.*, 27 F.C.C. 359 (1959).

91. Ibid., 384. One study of Los Angeles, for example, showed that 38 systems were currently in use. Yet using only frequency bands already allocated for private use, 807 systems were physically possible in the Los Angeles market. New York, by contrast, was the second most developed market, with 3 systems in operation.

92. Ibid., 387.

93. Ibid., 392.

94. Ibid., 409.

95. *Microwave Communications, Inc.*, "Initial Decision of Hearing Examiner," 18 F.C.C. 2d 979, 984, 1002 (1969).

96. Ibid., 1003.

97. Larry Kahaner, *On the Line* (New York: Warner Books, 1986), pp. 52–59.

98. *Microwave Communications, Inc.*, 18 F.C.C. 2d, 953, 971–972. Johnson vented his frustration with FCC practices in a lengthy article that, despite its cynicism, gives an interesting insider's perspective on regulation: Nicholas

Johnson and John J. Dystel, "A Day in the Life: The Federal Communications Commission," *The Yale Law Journal* 82 (July 1973): 1575–1634.

99. *Microwave Communications, Inc.*, 976–978.

100. *Specialized Common Carrier Services*, 29 F.F.C. 2d 870, Appendix A, pp. 943–945 (1971).

101. Ibid., p. 876.

102. Ibid., pp. 881–889.

103. Ibid., p. 915.

104. Defendants' Third Statement, p. 684; and, Steve Coll, *The Deal of the Century* (New York: Atheneum, 1986), pp. 20–24.

105. Plaintiff's Third Statement, pp. 473–480.

106. Defendants' Third Statement, p. 685.

107. Ibid., pp. 687–688; and Coll, *The Deal of the Century*, pp. 51–52.

108. According to Coll in *The Deal of the Century*, the Common Carrier Bureau chief, Walter Hinchman, learned the meaning of Execunet in a visit from an AT&T lobbyist. The lobbyist wanted to show Hinchman a new MCI service and gave Hinchman a series of numbers to dial on a touch-tone phone nearby. Hinchman obliged, heard the ring, and then listened to the Chicago weather a switched long-distance call between two nonprivate phones; pp. 83–85.

109. Richard E. Wiley, "Address Before the United States Independent Telephone Association," Dallas, Texas, October 16, 1975.

110. *MCI Telecommunications Corp.*, 60 F.C.C. 2d 25 (1976).

111. *MCI Telecommunications Corp. v. FCC*, 580 F.2d 590 (D.C. Cir.), cert. denied, 439 U.S. 980 (1978).

112. *Exchange Network Facilities for Interstate Access (ENFIA)*, 71 F.C.C. 2d 440 (1978).

113. Defendants' Third Statement, p. 847.

114. Plaintiff's Third Statement, p. 137.

115. *AT&T, Tariff FCC No. 250, TELPAK Service and Channels*, 38 F.C.C. 370 (1984).

116. *AT&T, Private Line Services, Series 5000 (TELPAK)*, 61 F.C.C. 2d 587 (1976).

117. AT&T, *Annual Report, 1970*, pp. 13–14.

118. Alvin von Auw, *Heritage and Destiny* (New York: Praeger, 1983), pp. 309–321.

119. Ibid., pp. 311–312; and, Temin, *The Fall of the Bell System*, pp. 62–63, 71.

120. Ibid., pp. 70–74.

121. Ibid., pp. 70–85; and, Coll, *The Deal of the Century*, Chapter 4.

122. Plaintiff's Third Statement, pp. 1103–1104.

123. Ibid., pp. 679–680.

124. Ibid., pp. 676–709.

125. John D. deButts, chairman, AT&T, "An Unusual Obligation," address to the annual convention of the National Association of Regulatory Utility Commissioners, Seattle, Washington, September 20, 1973. DeButts also told shareholders that "competition thus far and our projections of its consequences in the future have brought us to the conviction that its further extension would not be in the public interest and that, believing so, we have an obligation to the public to say so" ("Letter from the Chairman of the Board," September 19, 1973, in *Quarterly Report to the Shareholders*, September 1973).

126. To avoid the appearance of deregulating without considering its consequences, the FCC initiated an investigatory proceedings, docket no. 20003 (46 F.C.C. 2d 214 [April 10, 1974]). Although it induced some high-quality anal-

yses of economic consequences, this proceeding merely dissipated, with no clear resolve or impact.

127. *United States v. AT&T,* Complaint, November 20, 1974, in Sterling et al., *Decision to Divest,* pp. 4–18 and I-10–I-12; and, Coll, *The Deal of the Century,* pp. 63–72.

128. Davis Dyer, "AT&T at the Crossroads, December 1981," in *Telecommunications in Transition: Managing Business and Regulatory Change,* ed. Richard Vietor and Davis Dyer (Boston: HBS Case Services, 1986), pp. 37–38.

129. von Auw, *Heritage and Destiny,* pp. 331–335; and, Coll, *The Deal of the Century,* pp. 103–104, 130. Quoted in von Auw, *Heritage and Destiny,* p. 416.

131. Paul Lawrence and Davis Dyer, *Renewing American Industry* (New York: Free Press, 1983), pp. 215–137; and, Temin, "Getting Off the Hook," pp. 5a2–5a16.

132. Dyer, "AT&T at the Crossroads," pp. 34–35.

133. Quoted in R.Z. Manna, "In the Matter of Telecommunications Legislation," *Bell Telephone Magazine* 4 (1981): 6.

134. U.S. Congress, Senate, Commerce Committee, *Hearings on S.898,* "The Telecommunications Competition and Deregulation Act of 1981" (97th Cong., 1st Sess.), June 1981, p. 442.

135. Sterling et al., *Decision to Divest,* pp. 115–116.

136. *United States v. AT&T,* "Pretrial Brief for the United States," in Sterling et al., *Decision to Divest,* pp. 75–78.

137. *United States v. AT&T,* "Opinion" (denial of motion to dismiss), September 11, 1981, in Sterling et al., *Decision to Divest,* p. 859.

138. Quoted in Dyer, "AT&T at the Crossroads," p. 36.

139. W. Brooke Tunstall, *Disconnecting Parties* (New York: McGraw-Hill, 1985), p. 119.

140. *United States v. AT&T,* C.A. No. 82-0192, "Modification of Final Judgment," August 24, 1982, IId, in Sterling et al., *Decision to Divest,* p. 1295. The MFJ also constrained AT&T from offering any electronic publishing for seven years and, indefinitely, from purchasing the assets of any RBOC.

141. *In the Matter of MTS-WATS Market Structure* (Phase IO, C.C. Docket No. 78-72), "Notice of Proposed Rule Rulemaking," 67 F.C.C. 2d 757 (1978); for details of the plan's evolution, see John McGarrity, *Implementing Access Charges: Stakeholders and Options* (Cambridge, Mass.: Center for Information Policy Research, 1983).

142. The precise flows of these subsidies, as among various states and regions, was the product of a very complex pooling system. At divestiture, this system was reorganized as the National Exchange Carriers Association (NECA), with a series of new standards and procedures for managing the cross-subsidies. See V. Louise McCarren and Paul F. Levy, "The Mandatory Carrier Common Line Pool Should Go," *Public Utilities Fortnightly* 119 (March 19, 1987): 6, 23–27.

143. D. Davidson and R. Vietor, "Economics and Politics of Deregulation: The Issue of Telephone Access Charges," *Journal of Policy Analysis and Management* 5, no. 1 (Fall 1985): 10.

144. FCC, C.C. docket no. 78-72, Decision, 93 F.C.C. 2d 241 (1983).

145. NARUC, "Brief of Petitioner on Review of Orders of the Federal Communications Commission," October 12, 1983 (D.C.C., 1983).

146. U.S. Congress, Senate, Commerce Committee, and House, Energy and Commerce Committee, Joint Hearings Relating to *The Preservation of Universal Telephone Service* (98th Cong., 1st Sess.), May 1983.

147. John D. Dingell and Timothy E. Wirth, "The Great Phone Robbery," *Washington Post,* October 26, 1983, p. A27.

148. FCC, C.C. docket no. 78-72, "Letter from United States Transmission Systems, U.S. Telephone, Satellite Business Systems, EMX Telecom, MCI, GTE, Western Union, and Lexitel to Chairman Mark Fowler," October 4, 1983.

149. U.S. Congress, House, Committee on Energy and Commerce, *Report No. 98-479,* "Universal Telephone Service Preservation Act of 1983" (98th Cong., 1st Sess.), November 3, 1983.

150. FCC, C.C. docket no. 78-72, Memorandum Opinion and Order, February 15, 1984.

151. In 1985, with the FCC's access charge plan progressing very slowly, several of the Bell operating companies proposed alternative plans for restructuring access rates. The companies were allegedly alarmed by the growing trend toward bypass by their largest customers. In fact, these plans were designed to increase the BOC's control of long-distance customers through billing and service contact. None of these plans succeeded, in part due to the successful lobbying by a new coalition of consumer activists, AT&T, and MCI. For details, see Mark L. Lemler, *The FCC Access Charge Plan: The Debates Continue* (Cambridge, Mass.: Center for Information Policy Research, 1987).

152. Huber, *Geodesic Network,* p. 3.11. These aggregate figures, however, do not reflect substantial variations among segments. New entrants obviously targeted high-volume segments—large urban areas and business customers. Thus, AT&T reported in 1984 that competitors had captured 41 percent of business customers with expenditures greater than $150 per line per month and 25 percent of residential customers with expenditures greater than $100 per month; *In the Matter of Long-Run Regulation of AT&T's Basic Domestic Interstate Service,* C.C. docket no. 83-1147, "AT&T Filing," April 2, 1984.

153. National Telecommunications and Information Administration, U.S. Department of Commerce, "Telephone Competition and Deregulation: A Survey of the States," October 1986 (Washington, D.C.: NTIA, 1986).

154. *Third Computer Inquiry,* C.C. docket no. 85-229, "Notice of Proposed Rulemaking," July 26, 1985.

155. FCC, docket no. 85-229, "Report and Order," June 16, 1986; and, Mark Fowler, Albert Halprin, and James Schlichting, "Back to the Future: A Model for Telecommunications," *Federal Communications Law Journal* 38, no. 2 (August 1986): 188–193.

156. *Competitive Carrier Rulemaking,* 83 F.C.C. 2d 481 (1983).

157. FCC, docket no. 83-1147, "Notice of Inquiry," October 19, 1983.

158. *Guidelines for Dominant Carriers' MTS Rates and Rate Structure Plans,* C.C. docket no. 84-1235, "Comments of MCI Telecommunications Corporation," February 25, 1985.

159. Booz, Allen & Hamilton, Inc., "Prospects for Major Facilities-Based Other Common Carriers," prepared for GTE Corporation, March 18, 1985.

160. *The Wall Street Journal,* March 5 and 6, 1987.

161. See note 67.

162. *United States v. AT&T,* C.A. no. 82-0192, U.S. Department of Justice, Antitrust Division, "Report and Recommendations of the United States Concerning the Line of Business Restrictions Imposed on the Bell Operating Companies by the Modification of Final Judgment," February 5, 1987.

163. Peter D. Shapiro, "Public Policy as a Determinant of Market Structure: The Case of Specialized Common Carriers," in *Working Papers* (Cambridge, Mass.: Center for Information Policy Research, 1974).

164. National Telecommunications and Information Administration, U.S. Department of Commerce, *Assessing the Effects of Changing the AT&T Antitrust Consent Decree* (Washington, D.C.: NTIA, February 1987).

3

The Technological Imperative for Competition

Peter Huber

Telephone service in this country remains heavily regulated at both state and federal levels. Among the regulators are state public utility commissions, the Federal Communications Commission (FCC), and a federal district court enforcing the terms of the 1984 consent decree (the Modified Final Judgment, or MFJ) that led to the breakup of AT&T into eight corporate pieces. Regulation has heretofore been rationalized as necessary to control monopoly market power.

The history of the past two decades in U.S. telecommunications, however, has been one of steadily eroding monopoly. The principal driving force has been change in technology. Regulatory policy has adapted, sometimes quickly, sometimes only slowly and reluctantly. The question for regulators today is whether to accept and promote the ongoing evolution toward a fully competitive telephone network or to attempt to defend vestigial monopolies so as to advance other social policies.

Terminal Equipment

As recently as fifteen years ago, the idea of real competition in the markets for terminal equipment seemed farfetched. Handsets, private exchanges, modems, multiplexers, and other types of customer premise equipment (CPE) were part of "the network"—a monopoly largely owned and controlled by AT&T. Many were content to keep it that way. Equipment that was

Note: This chapter is adapted from *The Geodesic Network: 1987 Report on Competition in the Telephone Industry* (Washington, D.C.: U.S. Department of Justice, 1987).

located on customer premises—everything from handsets to mainframe computers—was provided only under lease, and then only grudgingly, with strict instructions that nothing was to be tampered with in any way. The real electronic brains stayed safely in the central office.

As recently as 1980, for example, total sales of private branch exchanges (PBXs) amounted to about one-quarter of central office equipment sales; the combined sales of private exchanges, key systems, modems, and other private switching and multiplexing equipment amounted to barely half of central office equipment. And most of the terminal equipment was provided by the telephone company, or by a second, almost equally monolithic business power, IBM. To be sure, a few competitors were chipping away at the edge of the citadel. But no one was talking seriously about robust competition in CPE markets. CPE was, in any event, unsophisticated and represented a comparatively small market.

Today, the center of gravity in telecommunications equipment sales has shifted decisively from carrier central offices to end-user private premises. Both AT&T and IBM used to be renters; both are now sellers. This alone has created millions of new owner-operators of telecommunications and computing systems. And end users have quickly grown accustomed to their new role; CPE markets have been booming. In 1981, PBXs served about 15 million telephone stations; today they serve somewhere between 25 and 30 million. New PBX-switched lines are now being added to the network as fast as new local exchange carrier (LEC) access lines. Private buyers now account for 80 percent of the purchases of satellite transmission services, 40 percent of the telephone switching market, 20 percent of microwave transmission equipment, and 20 percent of fiber optic cable and electronics. Computer penetration of the U.S. home market has grown at a rate unparalleled in the history of consumer electronics: from 340,000 units in 1981 to an estimated 12 million units by the end of 1985. All markets for CPE are characterized by robust competition.

How did we make this astonishingly rapid transition from a world of almost complete monopoly in terminal equipment to one of robust competition? The most important force was technology. The microchip revolution has pushed down the cost and vastly expanded the functionality of CPE. Today's desktop PBX offers power and functionality that twenty years ago could be offered, if at all, only through a large central office switch. The telephone handset, or at least all of its essential components, has been reduced to a single chip. And the phone-on-a-chip is rapidly migrating into the personal computer, the copier, the FAX machine, the electric meter, and the burglar alarm. Terminal equipment used to come with a single port to plug into the network. Today's PBXs, micro- and mainframe computers, and other intelligent terminals have many ports. This has opened enormous new competitive possibilities. A new business phone sold today, for example, is more likely to be connected to a private exchange than to the local public exchange.

Rapidly evolving technology, in short, has fundamentally changed the functions and capabilities of terminal equipment. More often than not, modern network "terminals" no longer terminate; they interconnect. Technology alone has rendered obsolete the old vision of a simple, unitary network, ending at the rotary-dial handset. Each new connection to a modern, multiport unit of CPE creates opportunities for two more.

This has brought into play a textbook axiom of antitrust theory. "Tying" a product sold in a competitive market to one provided in a noncompetitive market will successfully impede competition in the tied-product market only if the tied and tying products are used in fairly fixed proportions. That used to be so for LEC lines and CPE, in the days when CPE was truly "terminal" equipment, with a single port connecting to a single LEC line. But today, one network terminates where another begins. Even the lowliest LEC residential line serves an average of two phones and can serve many more. A modem, answering machine, automatic dialer, alarm company's digital dialer, and dozens of other pieces of equipment can be added incrementally to the same line, without the LEC's acquiescence or control. If the LEC line connects to a PBX or a computer, there is no limit at all to what else may hang on beyond. A tying strategy in this kind of market is doomed to fail.

The FCC recognized and responded to the new competitive possibilities in the early 1970s. Following inevitable court challenges, the commission segregated CPE markets and invited full competition. The commission set up a fairly straightforward registration program to ensure the compatability of equipment interconnecting with the network and then allowed competition to run its course. There are now hundreds of companies offering CPE of every description. By about 1982, two years before divestiture, AT&T had lost almost all its historic dominance of these markets. AT&T still was—and indeed still is—a powerful and innovative provider of CPE. But the old CPE monopoly is entirely gone, and not widely lamented, even if some customers are still frustrated by the division of responsibility between equipment manufacturers and network operators.

Private Line Service

Private lines provide dedicated, point-to-point connections, usually between two nodes owned by a single corporate customer. For many years AT&T was the overwhelmingly dominant —if not absolutely exclusive—provider of this type of telecommunications service. Once again, that is far from true today, in large measure because of changes wrought by technology.

Most important was the growth of the private exchange and the communicating mainframe computer. By the mid-1970s, large corporate and institutional users were increasingly often operating their own small telephone exchanges. The Bell System network had relied on five tiers of switching. Private exchanges began to creep onto the scene and began to form a sixth tier sitting between the end user and the local public exchange. An upstart company then thought to build a business around these developing private nodes. The plan was to offer cheap, point-to-point microwave links between the private nodes owned and operated by large private users. The company, of course, was Microwave Communications International—MCI. Following a pattern of development that has since been repeated elsewhere, the growth in privately controlled terminal equipment thus created the opportunity for more competition in the adjacent, complementary market for private line service. Regulation adjusted and competition rushed in to meet the new demand.

The transmission technology for providing point-to-point trunks in this market has continued to improve steadily. Microwave systems have developed in both range and reliability, and the FCC recently allocated a new spectrum suitable for short-haul microwave applications. Small-aperture satellite technology, capable of supporting duplex service, developed still more recently. Today, in areas served by the seven regional Bell operating companies (BOCs), the regionals currently provide just under 3 million circuits of in-use, private line capacity. The combined capacity (not fully used) of point-to-point private microwave and satellite systems in these same regions is almost twice as large. New companies like New York Teleport are now installing fiber optic networks to offer short-haul, private line service to major corporate users in urban areas. New York Teleport is the most advanced of these ventures but similar developments are under way in other cities. Overall, today there is fairly robust and growing competition in the provision of telecommunications trunks at the T-1 level or above.

Once again, regulation—at least at the federal level—has roughly kept pace with the new competitive possibilities. The first regulatory break came in 1959, with the FCC's *Above 890* decision, which permitted private users to install and operate their own dedicated end-to-end network links. Fifteen years later the commission extended its earlier initiative in approving MCI's point-to-point *Execunet* service. A pro-competitive regulatory policy thus intersected with rapidly developing technology to create new competitive opportunities. The market has since responded vigorously.

Switched Long-Distance Service

Until quite recently, only AT&T offered universal long-distance service in the United States. Even the smaller independent phone companies that served about a quarter of the nation's telephones routed their long-distance calls through AT&T's network. Today, by contrast, 70 percent of residential users are connected to local exchanges that offer "equal access" to competing long-distance carriers. Three large facilities-based carriers have their own voice switches and multiplexers scattered across the country. Whatever may happen to the companies that now own it, the long-distance wire in the ground is here to stay, as are the equal-access capabilities in the local exchange. It thus appears that one way or another, competition in long-distance markets will survive, notwithstanding AT&T's still considerable market power.

The critical technological capability here has been the switch (either analog or digital) operated under stored-program control. In the mid-1960s, AT&T introduced the electronic stored-program control (SPC) telephone exchange. In the following decade this technology gradually displaced the electromechanical exchange. And that transition, more than any other single factor, paved the way for full competition in the provision of switched-access long-distance service. It is the modern electronic switch that now supports the efficient, customer-controlled distribution of traffic among dozens of competing long-distance carriers. In short, advanced electronics provides the flexibility that makes competition affordable. Without stored-program control, the transition to equal access would have been prohibitively cumbersome and expensive.

Regulation has again evolved along with the technology, though somewhat more slowly. The first regulatory watershed was a decision by the D.C. Circuit Court of Appeals. In a creative reading of the FCC's earlier *Execunet* ruling (which had apparently addressed only private line service), the court concluded that MCI was also entitled to offer universal, switched long-distance service using the Bell System's local exchanges to collect and distribute traffic at the originating and terminating ends. But this ruling did not require equal access as we now know it. A few years later, the AT&T divestiture decree took the next regulatory step and established a timetable for the full implementation of equal-access capabilities in all Bell System exchanges served by electronic switches. Much of that transformation is now complete.

Electronic Information Services

Markets for electronic information services do not start with a clear history of monopoly, largely because the markets themselves have very little history to speak of. Nevertheless, until quite recently the common wisdom was that monopoly might well prove to be the destiny of these markets, too, at least absent the most careful regulatory and antitrust oversight. Accordingly, a 1980 FCC ruling barred AT&T from offering electronic information services through its local or long-distance phone companies. At the same time, the Department of Justice engaged in a titanic, interminable, and enormously costly antitrust suit against IBM, which appeared at the time to have all but untouchable power in computer markets. And in drafting the divestiture decree, the Justice Department took special pains to bar the seven regional Bell companies from providing any electronic "information services" at all. The federal district court reviewing the decree went one step further and also prohibited AT&T from offering any "electronic publishing" services for seven years.

The arguments raised against AT&T, IBM, and then the regional Bell companies sounded a common theme. All these companies had monopolized, or at least threatened to monopolize, the crucial gateways to electronic information services. They could manipulate the standards and interface specifications of their equipment to freeze out potential competitors. Or so the argument ran.

The arguments were perhaps plausible when they were made, but no objective observer could take them very seriously today. Once again, the critical development has been technological. Dramatic advances in the technology of microprocessors and electronic storage/retrieval systems have driven a very rapid dispersal of electronic intelligence out toward the end user. Sales of minicomputers and microcomputers have substantially eclipsed sales of mainframe computers. There are now hundreds of millions of intelligent chips installed in other devices and appliances. The center of the electronic information services market now lies well beyond the edges of the traditional telephone network; with only comparatively minor exceptions, it is still entirely outside it. Electronic processing, reformatting, storage, retrieval, and all the other elements of an electronic information service are now available in a robustly competitive marketplace. The technological imperative of the information age is the dispersion of electronic intelligence into computers, private exchanges, handsets, copying machines, utility meters, home appliances, automobiles, watches, electronic pacemakers, and children's toys. This dispersion is, by all appearances, irreversible; for every new capability added inside the public network, a dozen will be added outside.

Regulatory policy now appears to be inching toward this same conclusion. The FCC, which earlier had flatly barred regulated telephone companies (telcos) from offering any form of "enhanced services" through the public network, has recently begun to revise its policies. Its new regulatory philosophy is to encourage the provision of enhanced services both inside and outside the network. Nontelco providers of information services are to be guaranteed, whenever feasible, equal access to network facilities, on the same terms and conditions as the telcos themselves. The Department of Justice has recommended that the divestiture decree be modified to eliminate the absolute barrier against the provision of information services by the seven regional Bell companies. In September 1987, Judge Harold Greene, who oversees enforcement of the divestiture decree, agreed to "lift so much of the information services restriction [in the MFJ] as prevents the regional companies from constructing and operating a sophisticated network infrastructure that will make possible the transmission, on a massive scale, of information services originated by others, directly to the ultimate consumer."

The deployment of "open network architecture" (ONA) switch software in the local telephone exchanges will mark yet another major technological step toward a competitive network architecture. Open architecture will permit local exchange carriers to separate the elemental piece parts of what is now the bundled local exchange service—dial tone, switching, ring and busy tones, and answer supervision—and sell them individually, allowing independent providers to assemble their own customized package of services. In the future the LEC switch will do only that; other units—controlled by the LEC and by others as more or less equal providers—will provide all ancillary features. Very roughly speaking, ONA is the information service equivalent to equal access for long-distance voice carriers. The FCC has made the implementation of ONA a cornerstone of its new competitive policies.

The Local Exchange

The AT&T divestiture decree was based on the assumption that the trunking of high-density traffic can be handled in a competitive market but that collection and distribution of low-density traffic are a natural monopoly—except, that is, at the very lowest levels of the network, on the customer premises, where competition is possible once again. The local exchange does indeed seem to be the last and hardiest survivor in the competition's assault on traditional telephone monopolies. Nonetheless, important pro-competitive forces are evidently at work here too.

In the past hundred years, the basic low-density transmission technology—twisted copper wire—has not changed much, nor has its price. Fiber optic cable has slashed transmission costs for high-density applications, but the last mile of the network, where about half the transmission expense arises, carries mostly low-density traffic. In the past fifteen years, on the other hand, revolutionary developments in electronics have slashed the costs of network intelligence.

When switching is cheap and transmission comparatively expensive, as is increasingly true in the local network today, one does not build a network with local loops that run many miles to converge in a few bottleneck switches. It is more economical, instead, to move the network nodes and intelligence out toward the end user. Switches and other intelligent nodes proliferate; transmission is kept to a minimum.

It has always been true that equipment and transmission are substitutes at the margin: a national network can be assembled with a billion wires and one switch, or a billion switches and one wire, or any mix in between. In the past, however, the trade-off between switching and transmission occurred almost entirely inside the public network. But as the vital center of equipment markets has moved from the central office out to private premises, control over the transmission-switching trade-off has moved from carrier into private hands. In countless areas CPE now provides, at least at the margin, a direct substitute for local exchange, long-distance, and on-line information services. The entire national packet-switching industry has been built on the economies of trading off more packet switching for less long-haul transmission. The move from mainframe computers to microcomputers provides the ultimate form of bypass: no telecommunications at all. PBXs in corporate or shared-tenant hands deal with internal traffic (typically one-third of all calls) entirely on their own and cut the number of LEC lines needed for outside access by a factor of ten. The trick is to funnel a lot of traffic into a few lines—which PBX owners do especially eagerly whenever an LEC is foolish enough to tariff its lines at flat, unmeasured rates.

Judging by sales of CPE and central office equipment, that transition is already well under way. The private exchange is the critical piece of hardware that will move us down the slope, from a world of few switches and many long lines to a world of many switches and far fewer lines. And as we have seen, the installed base of PBXs has been growing rapidly.

The interexchange carriers have an especially strong incentive to promote competition in the local exchange by whatever means possible. The most ominous competitive threat to the interexchange carriers, and to AT&T in particular, lies in the BOCs. Today, AT&T returns about half of all its long-distance revenues straight back to the local telcos; access charges certainly dwarf profits. AT&T's strategic planners must be justifiably horrified at the thought that some day AT&T may find itself competing head to head with companies that control half of AT&T's own costs. It is inconceivable that AT&T and the other interexchange carriers will sit back complacently, paying inflated local-access charges at both originating and terminating ends of their networks, and simply hope that regulatory forces will protect them in the longer term from competition. But the public posture of interexchange on this issue is not, of course, likely to mirror the private one.

Nonetheless, the transition to competitive local exchange services is already under way in urban areas and in the provision of service to large institutional customers. Building wiring is fast becoming the principal means for concentrating business traffic at a first-tier private switch. Nationwide, there are in fact already more than 14,000 PBXs serving more than 400 lines each, and 40,000 smaller exchanges in the 100- to 400-line range. As the number of PBXs and large communicating computers grows, so will the opportunities for connecting them directly to each other. Networking is contagious. The more nodes there are in place, the easier it is to connect them up along alternative transmission paths.

Even in a world permeated with private nodes, there will still be the problem of short-haul transmission. Barring significant advances in parapsychology or telekinesis, one cannot have telecommunications without transmission. But as competitive local nodes proliferate, dependence on LEC lines does decline steadily. Transmission and intelligent nodes are highly substitutable, at least at the margin. As nodes proliferate, the troublesome "local loop" gets shorter. The number of loops needed declines rapidly. And the price one has to pay for local access, which depends principally on the number and length of local loops, declines rapidly.

In short, as dispersed electronic intelligence permeates the local exchange we can prepare to relive the MCI experience all over again, one tier lower down in the network. As the first hint of competition is allowed, resellers move in to exploit price dislocations created by cumbersome, unresponsive regulation. Competition in providing service at the nodes develops first, ahead of competition in supplying transmission. As more competitors arrive on the scene, regulators are pressed to make sure that transmission tariffs are properly priced. Cross-subsidies erode. And this opens up additional possibilities for competition in market niches that were previously unattractive because they were subsidized. Finally, competing transmission systems take shape to complete the competitive challenge. That, more or less, is a capsule summary of how MCI broke into the long-distance markets. That same dynamic is now under way at the level of the local exchange.

The regulatory response to the possibility of local exchange competition is now beginning to take shape. Federal regulatory policy appears to be solidifying in favor of competition. The Department of Justice has made competition in the local exchange a cornerstone of future policy. Specifically, the Justice Department has tied removal of current restrictions against the regional companies' entry into long-distance service markets to reciprocal deregulation of the local exchange. Most of the signs are that the FCC will similarly favor competition in the local exchange.

State regulatory policies are less clear. A few states have enacted new laws to encourage competition. But in most others, the policy—often supported more or less openly by the local telco—is to preserve an exclusive monopoly franchise for a single company. Competitors are either flatly barred or more subtly obstructed from either reselling LEC services or providing competing services over independent switches and lines.

Despite the still cloudy regulatory picture, the forces favoring competition are strong. Much of the critical technology needed to drive competition in the local exchange is already deregulated. While it is possible and likely that regulatory reaction will continue to obstruct competition in some regions for some time, the regulatory scaffold is not likely to survive the competitive assault. It is one thing to require "partitioning" of a PBX, another to monitor it; one thing to forbid competing intraLATA toll service, quite another to stop it; and one thing to announce a tax on "leaky" PBXs, quite another to collect it. Information is the most fluid of commodities, and telecommunications is the art of making it flow. Technology is moving us irresistibly toward dispersed network intelligence.

Forces Favoring Monopoly

What kinds of technological and economic developments might be more successful in moving us back toward significantly less competition in any segment of the telephone industry?

One may start with the markets for telephone equipment. Divestiture did not break up AT&T's equipment manufacturing operations; to the contrary, it permitted AT&T to enter new equipment businesses, which AT&T promptly did. Even in the United States, where divestiture has created the most hospitable possible environment for competition, equipment is still manufactured by a tiny nucleus of very large firms. The U.S. scene, of course, looks looser today than it did a few years ago, and so it is. But that is a reflection of expanding market boundaries (principally a big jump in U.S. sales by Bell Canada's manufacturing affiliate, Northern Telecom), not of manufacturer fragmentation. All the major new equipment suppliers selling in the United States are established manufacturers from other countries. And once the focus shifts, as it then must, to the world market, the trend toward consolidation among manufacturers is clear. There were twenty-six manufacturers in the world market when the first digital central office switching systems came into service in the late 1970s; by 1984, that number had dropped to eighteen. Building the next generations of optical superswitches and supercomputers will require enormous concentrations of capital, talent, and sustained engineering effort. As a result, the global telecommunications equipment market is likely to become dominated, in the next decade, by about a handful of players. Nonetheless, this prospect should not spell an end to competition. The geographic boundaries of the marketplace are expanding faster than the number of firms is contracting.

The future of competition in markets for long-distance services is under some shadow because of the revolution in fiber optics. Fiber optic transmission costs have been dropping at an astronomical rate—and there is no end to that decline in sight. In dealing with high-density traffic, the basic trade-off between switching and transmission is thus very different from the trade-off between these same factors at the level of the local exchange. Indeed, long-distance networks may well be moving toward an architecture that is highly centralized. One can visualize today a long-distance market built around something like a "Memphis switch." A Memphis switch is what Federal Express operates: you fly all your overnight mail from everywhere in the world to Memphis, you switch it once, then you fly it all back out again. Fiber optics makes it increasingly attractive to head in a similar direction with high-density phone traffic: once you have the traffic in one of your major pipes, you can haul to the end of the earth and back, and it may often make economic sense to do so.

Nevertheless, it does not seem likely that competition in long-distance markets will collapse under the economies of scale that fiber optic transmission now offers. Three interexchange carriers already have long-distance transmission facilities largely in place, with capital costs fully sunk. Whatever may happen to their corporate owners, these assets are not likely to disappear, nor is any one company likely to be so foolish as to attempt to acquire them all. Competition is therefore likely to survive here as well.

A return toward monopoly conditions in markets for electronic information services seems even less likely. IBM is still a powerful force, but it faces serious challenge from other U.S. and foreign firms. One cannot predict with any certainty what the unfolding drama with superconductivity might mean for economies of scale in computer manufacture. Nor is it possible to predict how the scale economies of photonics and optical transistors may play out in this market. But by all current indications, competition in the global computer markets is likely to survive and flourish, even while consolidation among major manufacturers continues. It is all but inconceivable that the established telephone companies could easily monopolize any significant segment of information service markets, even if regulatory restraints were significantly relaxed.

The prospects for truly robust competition in the local exchange remain the most uncertain. The LECs undoubtedly enjoy significant economies of scale in their operations, though it is far from clear that those economies are intrinsically larger, overall, than the economies of scale enjoyed by the dominant long-distance carrier, AT&T. Access loops are, however, expensive to install and significantly underutilized by most small residential and business users. Moreover, the introduction of integrated services digital network (ISDN) technology will more than double the carrying capacity of LEC wire. All of this suggests, of course, that there may be important economies in relying on a single monopolistic provider of short-haul access loops. There is, in addition, no likelihood that airwave transmission systems will soon offer an economically viable access substitute for twisted copper wire.

Competition undoubtedly is both technically and economically feasible in urban areas, especially in high-rise buildings. But it nonetheless seems likely that the smallest consumers, and in particular most residential users, will not soon be shopping for telephone service in a competitive market. Larger users, on the other hand, will enjoy access to an increasingly broad range of competitive transmission and switching alternatives. One cannot easily foresee a day when anyone will operate a phone system totally independently of LEC facilities and services. But the LEC monopoly is shrinking down toward its most primitive and least valuable element—wire in the ground. Wire is also quite readily amenable to regulation in both price and availability.

The most difficult reality for many policymakers is facing up to the need to abandon many historic cross-subsidies. Competition drives price toward cost, and as it does so it removes opportunities for cost shifting. One way or another, interexchange carriers will soon pay less for local access. Either access tariffs will go down or the carriers will use less access (with proportionately more network intelligence) and rely on bypass facilities where they can. Larger business users will likewise retreat from overpriced local services. Either the services to these users will come into line with costs or the users will rely very much more on their own equipment. Changes of this character are not generally welcome among those who must explain telephone pricing policies to constituents and voters. Such changes are nonetheless inevitable in an industry that is being driven toward competition at all levels by the rapid deployment of new technology.

A Response to the Pessimists

The MFJ was built on the assumption that the local exchange was and would remain a natural monopoly, while other segments of the telephone industry could be competitive. The legal assumptions and pronouncements have now acquired a life of their own. Removal of the "line of business" restrictions on the BOCs now hinges on the finding that the BOCs cannot use market power in the local exchange to impede competition in the new markets they seek to enter. This condition impels many firms that might face competition from the BOCs were the restrictions relaxed, to proclaim dogmatically that the local exchange was a natural monopoly in the beginning, remains so today, and will remain so forevermore. Indeed, the established interexchange carriers have real reason to back the rhetoric with action, and deliberately prolong and exaggerate their dependence on BOC and other LEC lines, solely to buttress their arguments for maintaining the regulatory status quo.

The technological forces are nonetheless powerful and have made nonsense of similar claims of "necessary" or "natural" monopoly several times before. For many decades the Bell System also insisted that competition was infeasible or uneconomical in markets for terminal equipment because of likely "harm to the network" from competitors' hardware. The Bell System likewise stoutly insisted, for several decades, that competition was unwise if not infeasible in markets for interexchange services, because of economies of scale and the "piece out" process, under which competing carriers would rely on the Bell network for hauling traffic over high-cost segments of the country (in mountainous regions, for example) while building competing links in low-cost regions. These and other equally wrongheaded prophecies are now all but forgotten. The network has easily survived the most robust and chaotic competition in CPE markets. And despite the fact that it remains the long-distance carrier of last resort, AT&T continues to operate profitably in competition with many other carriers.

What, then, is the final word on the "natural monopoly" of the local exchange? Until we try competition, we cannot know what is possible. As John Wenders remarks in his excellent book, *The Economics of Telecommunications,* "The only way to really find out is to open up the local market to competition, *allow the local companies to compete,* and see what emerges. If a monopoly emerges, at least we will know that it is a reasonably efficient monopoly, and that potential competition will be sitting on its doorstep ready to discipline it if it doesn't stay efficient."[1] Certainly the threat of competition would force the existing local monopolists, such as they are, to price services more efficiently. That many LECs today spend disproportionate time and effort preserving *statutory* protection of their supposedly "natural" monopoly suggests that competition in at least some segments of their market is not only possible but a very real threat. Natural monopoly may or may not be one real problem in the local exchange. But the unnatural monopoly created by statutory franchise is certainly another. Until we dissolve the de jure monopoly, argument about the robustness of the de facto monopoly is of little practical relevance.

Note

1. J.T. Wenders, *The Economics of Telecommunications: Theory and Policy* (Cambridge, Mass.: Ballinger, 1987), p. 249 (emphasis in the original).

Part Two
Competition in Existing Markets

4

Competition in Telecommunications: Moving Toward a New Era of Antitrust Scrutiny and Regulation

Kevin R. Sullivan

The debate over the existing restrictions on entry by the regional Bell operating companies (BOCs) into currently prohibited lines of business presages a new era in the regulation of the business of telecommunications.[1] These lines of business include provision of interexchange services and information services, as well as manufacturing of telecommunications equipment such as central office switches and terminal equipment.

We are on the cusp of dramatic changes in telecommunications, from both a technological and an antitrust/regulatory perspective. Technology is evolving at a seemingly ever quickening pace in primary telecommunications technologies such as fiber optics, cellular radio, and integrated services digital networks (ISDN) and in related technologies such as microcomputers and optical discs (CD/ROM). The pace of change in such technologies makes it almost impossible for the sophisticated executive to keep abreast of current developments, let alone see future technological events with any clarity or certainty.

Technological evolution makes it dangerous to try to predict the future strengths and weaknesses of industry participants. As this chapter demonstrates, it is equally difficult to predict the antitrust/regulatory changes in the telecommunications industry. It is useful, however, to at least postulate the antitrust/

Note: The author expresses his gratitude for the exceptional assistance of Patrick J. Pascarella and Francy Youngberg. Mr. Pascarella is an associate with Pillsbury, Madison & Sutro in Washington, D.C. Ms. Youngberg is a third-year student at the Harvard Law School.

regulatory issues that will drive the telecommunications industry in the next several years.

Background

Changes in how the antitrust laws and regulation help shape the telecommunications industry have kept up with and sometimes outpaced technological changes. As the decade of the 1980s began, a unified Bell System with AT&T in control dominated over 80 percent of local and long-distance service in this country and more than 80 percent of the telecommunications equipment industry as well. AT&T was subject to pervasive, if imperfect, regulation at the state and federal level and to a plethora of private antitrust lawsuits, as well as a major antitrust suit brought by the U.S. Department of Justice.

AT&T was divested in stunning fashion by a consent decree, entered in 1982 with the Department of Justice and approved by Judge Harold Greene of the Federal District Court of the District of Columbia. Approval and entry occurred only after extensive debate and public comment.[2] Divestiture took place in 1984. To AT&T, the promise of divestiture was a more stable antitrust/regulatory environment in which the corporation, stripped of its local Bell operating companies (BOCs), which were to provide "natural monopoly" services, would be able to compete aggressively in all competitive markets. As divestiture was about to take place, it was assumed that the BOCs would quietly tend to plain old telephone service (POTS). As proposed by the Justice Department and AT&T, the divestiture decree contemplated that the BOCs would be precluded from competitive services.[3] This restriction came under immediate and aggressive attack from state public utility companies (PUCs), the Federal Communications Commission (FCC), and Congress.

The District Court responded to this attack by ordering a number of changes to the AT&T decree. First, Judge Greene, in response to public criticism, modified the absolute nature of the restrictions by requiring the inclusion of a waiver provision, since there would be many businesses that the BOCs could enter without causing undue concern of competitive harm, and since industry changes—technological, economic, or regulatory— would justify removal of all restrictions at some point in the future.[4] Specifically, the AT&T decree allows the BOCs to obtain a waiver that would exempt them from *any* of the "line of business" restrictions upon showing that they could not use their monopoly power over local exchange service to impede competition in the market they sought to enter.[5]

Second, the court transformed a Justice Department suggestion into an order mandating that the decree's "line of business" restrictions be reviewed globally once every three years. That order is the subject of the pending proceeding concerning the Justice Department's "1987 Report and Recommendations," filed on February 2, 1987, and subject to ongoing review by the District Court.

Third, the court required a number of specific changes in the decree to allow the BOCs to provide certain competitive services. The BOCs were given the opportunity, for example, to provide Yellow Pages, even though such directories compete with others in the marketplace. In addition, the BOCs were given the right to provide (i.e., sell) customer premise equipment (CPE) in competition with other suppliers, such as AT&T. They were not, however, permitted to provide telecommunications equipment or to manufacture any equipment; the decree itself permitted the BOCs to provide such competitive services as cellular telephone service and Centrex services.

No sooner had the BOCs freed themselves from AT&T's dominance in January 1984 than they pushed for freedom from every other restraint on their ability to provide almost any product or service ranging from long-distance services to real estate sales. In the process, the BOCs challenged not only the restrictions in the divestiture decree (Section II [D] of the decree) but also the structural separation requirements of the FCC imposed in *Computer Inquiry II* (CI-II)[6] and the states' desire to monitor expenditures for regulated and unregulated services. In a nutshell, the BOCs as well as AT&T fell in love with the idea of "competition" in new markets.

More than 160 waivers allowing the BOCs to enter a variety of businesses were granted in the first three years after divestiture. However, the core restrictions on the provision of interexchange services and information services, as well as the restriction against manufacturing of telecommunications equipment and CPE, remain largely unchanged. As recently demonstrated by more than a hundred filings on the decree restrictions, the BOCs are exploring the outer limits of their competitive potential after the AT&T breakup.

In the meantime, action has continued on the regulatory front. The FCC has been dramatically pro-competitive, though not deregulatory, moving from the structural separation requirements of CI-II to the elimination of such requirements in return for the promise of equal access in information services in the form of open network architecture (ONA) and comparably efficient interconnection (CEI) in CI-III.[7] The FCC has also proposed a new cost-accounting system that seeks to attribute costs more directly and rely on outside audits and "benchmarks" to control cross-subsidization.

The states are equally active, though evolving in vastly different directions. Some are moving toward less and less regulation. Nebraska, for example, has developed a deregulatory environment allowing telephone companies pricing freedom, while preserving their franchise monopoly at the same time. Illinois, on the other hand, is moving to encourage more competition at the local level. Vermont is flirting with the use of a "social contract" that would stabilize prices for current basic services in exchange for more flexibility in future pricing of new services. This arrangement would enable the state to reduce rate-of-return oversight. Other states continue to favor comprehensive regulation, arguing that local exchange companies should contribute revenues from competitive ventures to subsidize the rate base and lower the cost of local service. Some states are even ready to define new services as part of basic telephone services and offer local exchange carriers (LECs) protection from competition.

Competitors of the established telephone companies in the emerging telecommunications industry—such as long-distance carriers, information service providers, and many equipment manufacturers—are seeking to maintain the current restrictions on the BOCs. These competitors see the BOCs, other LECs, and AT&T as major threats. They thus seek to (a) maintain the current restrictions on BOC entry into new businesses, (b) impose structural separation requirements on AT&T and the BOCs, and (c) require an equal access regime for interconnection to the network; they also seek (d) nondifferentiated pricing of access and (e) ubiquitous availability of customer information in the possession of the established carriers.

With this background of intertwined and potentially inconsistent regulatory and antitrust regimes, it is extremely difficult to predict the winners in this high-stakes competition over the rules of the game. It is also important to focus on the overriding legal issues that are likely to shape this industry in the next few years. This chapter will first review the current status of the competitive rules and then identify the likely new battleground as the telecommunications industry heads back into the mainstream of antitrust and regulatory analysis.

Many of the participants at this colloquium discussed the efficacy of the current restrictions prohibiting BOCs from participating in the core services and offered views as to whether those restrictions should be lifted, and if so, under what conditions. While that debate is significant, it is highly likely that at some point in the near future the restrictions on the BOCs, especially with respect to information services and manufacturing, will be loosened. In that case, the focus of the debate will shift to the antitrust and regulatory issues affecting BOC entry into these markets. Those issues are crucial not only for the BOCs but for all local exchange carriers, because the competitive concerns flow from a perceived control over a local exchange service monopoly.

Specifically, once the BOCs are allowed to enter new lines of business, such as information services and manufacturing, an analysis of the essential facilities doctrine and the state action doctrine in the context of anticipated antitrust action will be required. Indeed, after the colloquium, the information service restriction was partially lifted, allowing the BOCs to provide information transmission services (such as protocol conversion) and a variety of voice storage services.[8] As we look to the future, the essential facility doctrine, a fundamental theoretical underpinning of the AT&T case, will require closer scrutiny to see its effects on the new environment. In addition, the states' increasingly important role, particularly in emerging information service markets, will be reviewed in light of (a) recent decisions that have increased states' authority to prevent FCC preemption and (b) broadened state action doctrine.

The Essential Facilities Doctrine in a New Telecommunications Environment

If we assume that the BOCs are put in the same position as other LECs and allowed to provide information services subject to FCC and state regulation, the impact of antitrust regulation on the conduct of BOCs needs to be addressed. The restrictions in the AT&T decree, after all, flow from a perceived antitrust problem, and any analysis of the legal issues should start with a clear understanding of the AT&T theory.

The fundamental antitrust underpinning of the AT&T case was based on the theory that AT&T had abused its control of an essential facility—access to residential and business local telephone customers, that is, the local loop. Though the theory has sometimes been explained in terms of control over the central office switching system, most observers agree that the "bottleneck" evolves out of the control over the copper wire that links the business or residence to the central office.[9] The government, as well as private plaintiffs, argued that AT&T had abused its control of the "essential facility" consisting of local access in order to impede competition in related markets (such as long-distance services) and the procurement of telecommunications equipment.[10] The theory was of fairly traditional antitrust origin, evolving out of the terminal railroad case of 1912.[11] In that case, a number of railroads controlled the only bridge into St. Louis and denied access to the bridge to nonmember railroads. The Supreme Court decreed the bridge a bottleneck for east-west traffic and declared that the agreement to preclude access to nonmembers was an unreasonable restraint of trade in violation of the Sherman Act (15 U.S.C. § 1).

The relief ordered was to require the consortium to provide *reasonable* access to other users to the extent that capacity was available. This case, though it was decided based on a conspiracy theory under Section 1 of the Sherman Act, led to a series of both single-firm and multifirm cases,[12] including the telecommunications cases brought against AT&T by MCI and Sprint.[13]

The elements necessary to establish liability under the essential facilities doctrine are:

> (1) control of an essential facility by a monopolist; (2) a competitor's inability practically or reasonably to duplicate the essential facility; (3) the denial of the use of the facility to a competitor; and (4) the feasibility of providing the facility.[14]

In cases involving the control or allocation of a bottleneck facility, courts have consistently held that the party controlling the facility "must use *reasonable selection criteria* in allocating the unique commodity,"[15] and *if feasible,* must give competitors reasonable access to the facility.[16]

In *Southern Pacific,* plaintiffs alleged that AT&T monopolized the market for intercity business telecommunications services in violation of the Sherman Act. The plaintiffs' claims were based on the theory that the BOCs' local distribution facilities were "essential facilities" and that denial of access or restricted access was prohibited by the antitrust laws. While recognizing that the antitrust laws do prohibit AT&T from *unreasonably and discriminatorily restricting access* to these essential facilities,"[17] the Court of Appeals went on to observe that "[a]bsolute equality of access to essential facilities, however, is not mandated by the antitrust laws."[18] As the court stated, "[A]ccess to essential facilities [need] be afforded to competitors upon such just and reasonable terms and regulations as will, in respect of use, character and cost of service, place every such company upon as nearly an equal plane as may be with respect to expenses and charges as that occupied by the proprietary companies."[19] However, added the court, "the antitrust laws do not require that an essential facility be shared if such sharing would be impractical or would inhibit the defendant's ability to serve its customers adequately."[20] Technical infeasibility is, therefore, a defense to an essential facilities claim.[21]

The essential facilities doctrine and the issue of whether a monopolist has a duty to aid its competitors were discussed in two recent cases.[22] *Aspen* and *Western Union* both involved the severing of a cooperative relationship between a firm that had monopoly power and a competing firm.

In the *Aspen* case, the Supreme Court held that the owner of three of four major ski areas in Aspen, Colorado, had no valid business reason for discontinuing its participation in an "all-Aspen" lift ticket that allowed skiers to use all four mountains. Refusal of the owner of the three areas to cooperate with its smaller competitor violated Section 2 of the Sherman Act (15 U.S.C.A. § 2), which prohibits monopolization or attempts to monopolize.

In contrast, the *Western Union* case involved a monopolist *exiting* from the market of telex terminal equipment. Originally, Western Union provided its salespeople with a list of independent vendors of competing equipment and encouraged its sales force to sell competitors' equipment, but later, when Western Union determined that its liquidation was going too slowly, the company withdrew the list of competitors, thereby forcing the plaintiff to recruit its own sales force. The Seventh Circuit held that Western Union did not violate Section 2 of the Sherman Antitrust Act.

Judge Richard Posner, writing for the court in *Western Union* emphasized the recent shift of antitrust policy from protection of competition as a process of rivalry to protection of competition as a means of promoting economic efficiency. Therefore, maintained Posner, the lawful monopolist should be free to compete like everyone else to prevent the antitrust laws from becoming a safety net for inefficient competitors. It is clear that a firm with lawful monopoly power has no general duty to help its competitors, either by holding a price umbrella over their heads or by pulling its competitive punches.[23]

A firm that controls a facility essential to its competitors may be guilty of monopolization, however, if it refuses to allow them access to the facility.[24] In the *Otter Tail Power* case, a wholesale supplier of electricity refused to supply electrical power to a power system that competed with it in the retail electrical power market and had no other supply. Similarly, Aspen Ski Co., by opting out of the "all-Aspen" ski ticket, effectively foreclosed competition from Highlands because the latter, having only one mountain, could not compete without offering its customers access to the other three ski mountains.

Western Union is distinguishable from *Aspen* because it is not an essential facility case. It would have been, had it refused to supply telex service to a customer who got his or her terminal from Olympia, or had Olympia been a competing supplier of telex service that, like the specialized common carriers in the long-distance telephone market, depended on the owner of the local exchanges to complete its service. Olympia had the alternative to hire its own sales force, whereas it would have been impossible for Highlands to acquire more mountains. Moreover, the "all-Aspen" ticket was common in the ski resort industry, whereas Western Union's action in providing a list of its competitors to its sales force was unusual and not based on any contractual obligation.

Two issues arise out of the essential facilities doctrine as it may apply to an LEC's provision of information services: (a) the effects of bypass on the status of BOCs as holders of essential facilities, and, assuming the continued applicability of the doctrine, (b) whether the AT&T decree-imposed obligations will comport with FCC or state regulatory requirements.

Does Bypass Limit the Applicability of the Essential Facilities Doctrine?

In the past four years, the BOCs and other LECs, have argued that they no longer control an essential facility in local telecommunications. Numerous studies by the FCC as well as by other research groups have addressed this issue.[25] In addition, the Huber report, prepared by another participant in this colloquium, spends a great deal of time assessing the factual issue of whether LECs possess control over any essential facility.[26] The fact that much time has been spent on this issue has not, however, resulted in a final answer to the question.

For a number of users, the Huber report demonstrates that there are considerable alternatives to local exchange telecommunications. Indeed, with respect to the CPE market, the report notes the following:

> Though they [the BOCs] can undoubtedly use the market power they still possess to erect nuisance barricades along the way, the RBOCs [regional Bell operating companies] can do little enough to stop this outward migration of network intelligence. . . . But even if regulatory process were to collapse entirely, the RBOCs would succeed in sucking only a comparatively small part of the CPE market back to their exclusive embrace.

More strikingly, with respect to information services, the report concludes that "the Bell System, old or new, could no more monopolize today's electronic information service markets by integrating features into the network than it could monopolize wheat farming by growing grain along telephone pole rights of way."[27]

Despite Huber's conclusions that the structure of today's network would indicate that there is little, if any, monopoly power left at the local level, his report is based more on a structural, theoretical paradigm predicting what will happen than on hard data depicting what has occurred. At this point, the hard data are sparse, and detailed documentation of actual bypass has not yet been produced. However, such data should be forthcoming in the next few years, and they suggest that for particular types of local users, such as moderate to large business customers, LECs may be found *not* to possess an essential facility.

Assuming that the BOCs obtain permission to enter information service markets or manufacturing and equipment markets, the issue of whether bypass is sufficiently prevalent to dissipate the essential facility will be significant in any future antitrust case. The mere fact that prior cases, including the series of AT&T cases during the 1970s and 1980s, were based on a conclusion that the BOCs possessed essential facilities at the local level will not preclude a relitigation of this same question in the years ahead. At the same time, it is unlikely that bypass will amount to a complete replication of the existing local network. Thus, if 20 percent of local customers have access to bypass technologies, a question likely to be litigated is whether that fact causes the local network to cease being an essential facility. As the controlled percentage of the local loop goes down, the applicability of the doctrine has less and less force.

What Does Reasonable Access Mean?

Assuming the continued applicability of the doctrine, the type of access to local facilities that will be required under an antitrust standard may differ from that under an FCC standard or an AT&T decree standard. In developing or defending any future antitrust case, the potential relief one is entitled to obtain will determine to some extent whether liability exists in the first instance. The relief granted in the requirement of equal access in the AT&T decree contrasts substantially with the less demanding reasonable access standard of antitrust case law.[28] The AT&T decree used the phrase *equal access* in conjunction with, and as defining, nondiscrimination. The decree imposes what would appear to be an absolute requirement of equality between AT&T and other long-distance carriers in the physical interconnection at the switch and in terms of the charges for the interconnection. That same standard appears to apply as well to BOC access to the provision of information services. Whether such a standard is compatible with FCC rules is also material. The FCC has shown a willingness (a) to give more weight to efficiencies in the provision of certain ancillary services to local exchange service and (b) not to require total equality in price, even though it may continue to impose equality in interconnection.[29]

When and if BOCs are permitted to provide information services, arguments concerning the type of access required are likely to pull any future antitrust cases in two opposite directions. On the one hand, plaintiffs in these actions who would compete with the BOCs in the provision of the ancillary competitive service will argue that the requirements of the AT&T decree and those imposed by the FCC in regulatory proceedings require absolute equality. On the other hand, the BOCs, or for that matter other LECs, are likely to maintain that the FCC requirements in certain circumstances have been lessened and the requirements of the AT&T decree, to the extent that they continue to apply, are unrealistic. Local carriers will argue that they should be subject only to the standard of traditional anti-trust cases, which is reasonable access, and that absolute equality is neither appropriate nor required. A problem that all local carriers will face in any of these disputes is that AT&T, when it controlled the BOCs, argued that equal interconnection for long-distance competitors was an impossibility. As we know, what was impossible in 1981 is a present-day technical reality for a majority of end-office switches in this country. Long-distance equal access, however, was only a ripple in the regulatory pond compared with the CEI/ONA tsunami about to hit the shore.

Under the FCC's CEI/ONA orders, the BOCs are required to furnish to information service providers interconnection to local exchange systems on an equal access, or "fully technical equal," basis. What exactly this requirement means the FCC has yet to say (possibly because it does not know exactly how the BOCs will provide the requisite access). The BOCs, however, are not left completely without guidance.

For example, in its "First Reconsideration Order," the FCC explained that while equal access requires something more than "rough comparability," it "does not demand impossible or grossly inefficient over-engineering of the network so that absolute equality is always achieved."[30] Thus, problems such as un-avoidable minimal distortion—resulting, for example, from amplification—that does not affect end-user perception or use would be acceptable under the equal access standard.[31] Also acceptable is the BOC's use of different interconnection facilities for its own information service operations. A carrier that chooses to provide interconnection to competitors through different facilities, however, will bear a somewhat heavier burden of demonstrating that the FCC's equal access requirement is met.[32]

As mentioned above, a key factor at which the FCC will look in considering CEI/ONA plans will be the equality of the services as perceived by the end users. Probably a bigger factor, however, will be the equality as perceived by competing information service providers, since they will be charged because of practical constraints on FCC staffing to police the interconnection. Non-BOC telecommunications providers have always been active and vocal participants with regard to the FCC's regulatory oversight of the BOCs. In the case of CEI/ONA, the FCC has sought to channel and use this propensity toward helpfulness. In formulating its ONA plan, each BOC is required first to consult with industry representatives and then to identify in its ONA plan the procedures and industry participation involved in its development.[33] Thus, the FCC's approach seems to be one of offering initial guidance and then allowing the industry to work out an acceptable and feasible solution, all subject to final approval by the FCC.

Determining type and quality of service, however, is only half the problem. What price the BOCs will charge for that service is also at issue. "How much?" is the first question most consumers will ask and is the question future information service providers are asking right now. In response, the FCC has provided somewhat more detailed guidance than it has with regard to the equal access standard.

Recognizing that co-location or integration can result in true efficiencies, the FCC has refused to adopt a general cost-averaging approach, which, the FCC fears, could mitigate these efficiencies.[34] Accordingly, some components of the BOCs' CEI or ONA plans will be cost based, while others will be computed on an average cost basis. For example, the FCC has determined that price averaging would be inappropriate for the distance-sensitive costs of connecting an information service provider to a carrier's interconnection facility or for traffic concentration equipment located on the information service provider's premises.[35] Interconnection cost, however, will be priced on an average cost basis regardless of whether the BOC affiliate connects in the same fashion as its competitors.[36]

In developing this pricing system, the FCC focused on a number of factors, including (a) the likelihood that actual efficiencies might be lost through price averaging and (b) whether the BOC has ultimate control over the cost of the facilities being provided. With respect to interconnection charges, the FCC observed that the BOCs would have quite "substantial control" over the costs and therefore would have no incentive to design and install the most economic type of facilities for their competitors if cost-based pricing were used. On the other hand, if distance-sensitive interconnection costs were averaged, information service competitors would have no incentive to locate in areas in which these costs were minimized and consumers would be deprived of benefits from the true economic savings in the form of transmission efficiencies. Applying this type of reasoning to each generic component of the BOC services to be offered, the FCC established the following pricing guidelines:

1. Interconnection costs will be priced on an average cost basis;
2. Distance-sensitive transmission services will be priced on a cost basis;
3. Traffic concentration equipment located at a BOC's interconnection facility will be priced on an average cost basis, with concentration equipment located on the provider's premises priced on a cost basis; and
4. Network usage costs will be provided under tariffs.

If implemented, these criteria should resolve much of the present uncertainty over CEI/ONA. Final resolution of this issue must, however, await the FCC's decision approving or rejecting the formal ONA plans filed by the BOCs on February 1, 1987. These plans have been subject to vehement opposition, and are not likely to be approved or rejected before the end of 1988. Finally, regardless of the resolution of ONA/CEI at the FCC, it is almost certain that the conclusion will be different from either AT&T decree requirements or general antitrust requirements.

The topography of the antitrust/regulatory battle is very likely to change in the next few years and may move the telecommunications industry toward traditional antitrust analysis. This movement will be aided not only by a reapplication of the essential facilities doctrine, as I have just discussed, but also by substantial changes in what is known in antitrust circles as the state action defense.

State Action Can Immunize Otherwise Anticompetitive Conduct by Local Exchange Companies

One need only look at the state action defense to antitrust liability to obtain an example of how legal/regulatory changes can affect perceptions and possibilities in antitrust suits, particularly as such suits arise in the telecommunications industry. When the AT&T case was being litigated in the 1970s and early

1980s, the trend in Supreme Court decisions generally was to restrict antitrust immunity in regulated industries. AT&T itself was unsuccessful in arguing that the pervasive regulatory scheme at the state and federal levels preempted antitrust liability.[37]

The state action defense was subject to similar retrenchment during that era. The defense actually originated in 1943, when the Supreme Court ruled that states could impose restraints on competition,[38] thereby upholding a California statute under which the state director of agriculture supervised a private cartel of raisin producers for the avowed purpose of stabilizing prices and preventing "economic waste."[39] Under *Parker* and subsequent cases, antitrust immunity existed for regulated industries if they could show (a) that the challenged anticompetitive conduct was a type "clearly articulated and affirmatively expressed as state policy" and (b) that the state actively supervised any anticompetitive conduct.[40]

The state action defense has meandered, somewhat aimlessly, over the years. One case suggested that anticompetitive conduct could escape antitrust scrutiny only if "compelled" by the state.[41] Despite this holding, questions persisted whether private action taken pursuant to state policy was entitled to immunity.

A recent Supreme Court decision, the *Southern Motor Carriers Rate Conference* case, has significantly broadened the use of the state action doctrine.[42] Legislatures of North Carolina, Georgia, and Tennessee expressly permitted motor carriers to submit collective rate proposals to public service commissions that had the authority to accept, reject, or modify the recommendations. Based on the permissive policy of the states, the Supreme Court held that the collective rate-making activities of rate bureaus were entitled to Sherman Act immunity under the state action doctrine. The *Midcal* two-prong test[43] was applied. The first prong was met because the state clearly articulated its intent to adopt a permissive policy. The second prong, requiring state supervision of private anticompetitive conduct, was also met because the state public service commission actively supervised the collective rate-making activities of the rate bureaus. This case was significant because it established that a state policy that merely "*permits*" rather than compels anticompetitive conduct otherwise subject to the antitrust laws can qualify to protect regulated industries from suit even in situations in

which that conduct constitutes private action, so long as there is
state supervision.

One limitation on state action immunity is that it has been
read as being limited only to the primary activity subject to
regulation. The Supreme Court has ruled, for example, that
while state action could displace competition in the provision of
electricity, it did not extend to permitting anticompetitive conduct in the provision of light bulbs.[44] By analogy, this decision
would raise questions as to whether a generalized state regulatory policy could immunize LECs' provision of services ancillary
to basic telephone service. It is not difficult to imagine a situation in which some "information services" are construed to be
within the umbrella of immunity protected by state action,
whereas other services are subject to antitrust scrutiny. Voice-
storage services or "Hello Yellow" services (i.e., electronic Yellow Pages) would seem to be the type of information services
that might well fit within the scope of a state policy to displace
competition with regulation of telephone service. This analysis
has been bolstered by two recent cases wherein coin-operated
telephone service and remote alarm systems have been included
under the umbrella of state regulation and wherein the state
action defense has been used to immunize allegedly anticompetitive conduct.[45] Other "information services" might not fit
within that same policy. Data processing services and traditional information retrieval services come immediately to mind.

Southern Motor Carriers creates some very interesting issues
if the information service restriction on the BOCs is lifted or
if BOCs are permitted to provide intrastate, interLATA long-
distance service. The AT&T decree's current prohibitions
against providing these services and its insistence on nondiscrimination under the mantle of equal access are not necessarily
consistent with the desires of state PUC regulators. The regulators may desire to have these services provided on a monopoly
basis. Again, there is a potential for conflict. These state regulators now have more interest in, and probably more authority
over, services that are wholly intrastate. That is because those
same federalist principles which seem to have driven the Supreme Court's dramatic infusion of new life into the state action
doctrine in *Southern Motor Carriers* also seem to have affected
the Court's recent decision substantially limiting the FCC's authority to preempt state regulators. In *Louisiana Public Service
Commission v. F.C.C.* (1986) (106 S.Ct. 1890), the Supreme

Court rejected an FCC effort to preempt state regulation of the intrastate portion of depreciation schedules, holding that the FCC did not have jurisdiction with respect to charges, classifications, and practices in connection with intrastate telecommunications services.[46]

The increased applicability of the state action defense, along with the resurrection of states' rights to regulate telecommunications services, provides the potential for geometrically increasing confusion in today's already convoluted antitrust/regulatory regime. Ever since the divestiture of the BOCs took effect in 1984, observers have noticed that the "regulatory" nature of the AT&T decree presented a substantial possibility of conflict with FCC regulation. If I am correct about the potential applicability of the state action defense, the potential for conflict is increased roughly by a factor of fifty. Under a new regime of active involvement by fifty states, the FCC, and the antitrust laws, future debates over the efficacy and status of competition in telecommunications will reach a new level of complexity.

Conclusion

Changes in the AT&T decree "line of business" restrictions are likely to create new issues in the interplay of antitrust law and regulation at the state and federal levels. New questions will arise in terms of what "reasonable access" means or what "equal access" means under varied regulatory schemes. Moreover, it appears that there may be a resurgence of the state action defense at the same time as the states are being given more authority to limit federal preemption. Thus, just as we seem to be cementing a national environment of competition in telecommunications, the pendulum is set to swing back to state imposition of "beneficent" monopolization under a regulatory regime.

Notes

1. In telecommunications, the word *regulation* is generally understood to mean rate-of-return regulation by state public utility commissions (PUCs) or the Federal Communications Commission (FCC). In its broader context, of course, *regulation* encompasses all forms of PUC and FCC control over the telephone business and also includes the proscriptions of the antitrust laws.

2. *United States v. AT&T* (D.D.C. 1982) 552 F.Supp. 131 *affirmed sub nom. Maryland v. United States* (1983) 392 U.S. 659.

3. See Section II(D) of the AT&T decree.

4. See *United States v. AT&T,* supra at 194.

5. Section VIII(C) of the AT&T decree.

6. Petition of BellSouth Corporation and the Ameritech Operating Companies for Expedited Relief from and Limited Waiver of Computer II Structural Separation Requirements (filed November 7, 1985); Supplemental Data of Ameritech Operating Companies (filed November 12, 1985); Petition of Bell Atlantic Telephone Companies for Rulemaking to Permit the BOCs to Provide CPE Without Structural Separation (filed November 15, 1985); Petition of Southwestern Bell Corporation for Expedited Relief from and Limited Waiver of Computer II Structural Separation Requirements (filed November 20, 1985).

7. In Re Matters of Amendment of Sections 64.702 of the Commission's Rules and Regulations (Third Computer Inquiry) and Policy and Rules Concerning Rates for Competitive Common Carrier Services and Facilities Authorizations Thereof, Communications Protocols Under Section 64.702 of the Commission's Rules and Regulations, C.C. docket no. 85-229, Report & Order, FCC 86-252 (released June 16, 1986), Paragraph 212 ("Computer III Order"); Amendment of Sections 64.702 of the Commission's Rules and Regulations (Third Computer Inquiry), 2 FCC Rcd. 3035 (1987) ("Phase I Reconsideration Order").

8. *United States v. AT&T,* supra (Memorandum Decision March 7, 1988).

9. See, for example, *United States v. AT&T,* supra, 524 F.Supp. 1336.

10. *United States v. AT&T,* supra note 9, at 1336.

11. *United States v. Terminal Railroad Association of St. Louis* (1912) 224 U.S. 383, 411.

12. For example, *Otter Tail Power Co. v. United States* (1973) 410 U.S. 366 (control over ability to wheel electrical power at wholesale); *Hecht v. Pro Football Inc.* (D.C.Cir. 1977) 550 F.2d 982, 992 (control over the only professional football stadium in a jurisdiction is essential facility where implication of the facility is economically infeasible).

13. *MCI Communications v. American Tel. & Tel. Co.* (7 Cir. 1983) 708 F.2d 1081, 1132–33; *Southern Pacific Communications Co. v. American Tel. & Tel. Co.,* 740 F.2d 980.

14. *MCI Communications v. American Tel. & Tel. Co.,* supra; see also, *Otter Tail Power Co. v. United States* (1973) 410 U.S. 366; *Hecht,* supra, 570 F.2d at 982, 992–93.

15. *Central Chemical Corp. v. Agrico Chemical Co.* (D.Md. 1982) 531 F.Supp. 533, 552, affirmed (4th Cir. 1985) 1985-2 CCH Trade Cas. § 66, 753 (emphasis supplied).

16. *United States v. Terminal Railroad Association of St. Louis,* supra, 224 U.S. at 411; *Southern Pacific,* supra, 740 F.2d at 1008; *Byars v. Bluff City News Co., Inc.* (6 Cir. 1979) 609 F.2d 843, 856.

17. Ibid., 740 F.2d at 1008.

18. Ibid., at 1009. Equal access is, of course, a requirement of the divestiture agreement ending the AT&T case—*United States v. Western Electric Co., Inc.* (D.D.C. 1982) 552 F.Supp. 131, 227, affirmed sub nom. *Maryland v. United States* (1983) 460 U.S. 1001. Equal access also is required by FCC regulation and, in the name of comparably efficient interconnection, is a cornerstone of its Computer III Ruling (In Re Amendment of Sections 64.702 of the Commission's Rules and Regulations [Third Computer Inquiry], C.C. docket no. 85-229 [May 15, 1986]).

19. Ibid., quoting *Terminal Railroad,* supra, 224 U.S. at 411.

20. Ibid., quoting *Hecht,* supra, 570 F.2d at 92–93; emphasis supplied; *MCI Communications,* supra, 708 F.2d at 1133.

21. *United States v. AT&T,* supra note 9, 1360–61 (in considering suffi-

ciency of access to essential facilities, a court may consider problems of feasibility and practicability).

22. *Aspen Skiing Company v. Aspen Highlands Skiing Corporation* (1985) 105 S.Ct. 2847. *Olympia Equipment Leasing Co. v. Western Union Telegraph Company* (7 Cir. 1986) 797 F.2d 370, rehearing denied; 802 F.2d 217.

23. *Olympia Equipment Leasing Company v. Western Union Telegraph Co.* (7 Cir. 1986) 797 F.2d at 375.

24. *Otter Tail Power Co.*, Note 12 above.

25. For example, Telephone Communications: Bypass of the Local Telephone Companies (August 1986) Government Accounting Office; MTS and WATS Market Structure, Phase I, Third Report and Order 93 FCC 2d 241 (1983); MTS and WATS Market Structure and Public Notice Seeking Data, Information and Studies Related to Bypass of the Public Switched Network, (Docket no. 78-72, Phase I, FCC 84-635/rel. January 18, 1985); Touche Ross, *Local Exchange Bypass*, Summary of Touche Ross Studies (March 1986).

26. Peter W. Huber, *The Geodesic Network: 1987 Report on Competition in the Telephone Industry* (Washington, D.C.: U.S. Department of Justice, 1987), at 1.14.

27. Ibid., at 1.30.

28. It can be argued that the AT&T decree requirement of "equal access" is more stringent than would otherwise be required under the essential facilities doctrine. This is not unusual, however, because it is often the case that for "remedial purposes" injunctive relief in government antitrust cases goes beyond the strict requirements of the Sherman Act.

29. Computer III Order, Note 7 above, at ¶ 34.

30. Phase I Reconsideration Order at ¶ 92.

31. Ibid.

32. Computer III Order at ¶ 218.

33. This requirement has led to a series of "open" industry forums through the auspices of Bell Corps.

34. Computer III Order at ¶ 172; Phase I Reconsideration Order at ¶ 125.

35. Ibid., at ¶¶ 173–176, 180–182.

36. Ibid., at ¶¶ 177–179.

37. *United States v. AT&T*. (D.D.C. 1978) 461 F.Supp. 1314, 1320–21.

38. *Parker v. Brown* (1943) 317 U.S. 341.

39. *Parker*, Note 31 above, at 346.

40. *California Retail Liquor Dealers Association v. Midcal Aluminum, Inc.* (1980) 445 U.S. 97, 105; *City of Lafayette v. Louisiana Power and Light Co.* (1978) 435 U.S. 389, 410. The second requirement, active supervision by the state, is easily met in telecommunications (e.g., *Coin Call Inc. v. Southern Bell Tel. & Tel. Co.* [N.D.Ga. 1986] 636 F.Supp. 608, 613).

41. See, e.g., *Goldfarb v. Virginia State Bar* (1975) 421 U.S. 773, 790 (threshold question for state action is "*whether the activity is required by the State acting as sovereign*").

42. *Southern Motor Carriers Rate Conference v. U.S.* (1985) 105 S.Ct. 1721.

43. *Midcal Aluminum, Inc.*, Note 33 above, at 102.

44. *Cantor v. Detroit Edison Co.* (1976) 428 U.S. 579.

45. *Coin Call Inc. v. Southern Bell Tel. & Tel. Co.*, Note 33 above; *Sonitrol of Fresno, Inc. v. American Tel. & Tel. Co.* (D.D.C. 1986) 629 F.Supp. 1089.

46. *Louisiana Public Service Commission* was followed by a case in the U.S. Court of Appeals limiting the FCC's efforts to preempt state regulation of purely intrastate radio common carrier services (*People of State of Cal. v. F.C.C.* [D.C.Cir. 1986] 798 F.2d 1515).

5

United States v. AT&T: An Interim Assessment

Roger G. Noll and Bruce M. Owen

The purpose of this chapter is to assess the current direction and likely future development of the telephone industry as a result of the cataclysmic events of the early 1980s. We focus on two of the three major segments of the industry: telephone equipment and long-distance service. The third major segment, local exchange service, is treated only in passing.

The chapter is organized as follows: We first briefly review the major postwar developments in the industry that culminated in the Federal Communications Commission's (FCC's) decisions to experiment with competition in the provision of telephone equipment and long-distance service. We next describe the rationale that guided the Department of Justice in bringing its famous 1974 antitrust action against AT&T, and we then discuss the relationship between the structural changes embodied in the 1982 settlement agreement and the goals of the lawsuit. The balance of the chapter describes events in the years following the lawsuit's settlement, the extent to which the original goals of the Department of Justice have been achieved, and the challenges that lie ahead.

Note: The authors represent Stanford University and Economists Incorporated, respectively. Both authors were associated with the prosecution of *United States v. AT&T* and, in the past, have been consultants to competitors of AT&T and of the regional operating companies created by the decree. The authors are grateful to Charles Jackson and Jerry Hausman for especially helpful comments on an earlier draft and to Thomas Spavins for useful observations as our discussant. Roger Noll gratefully acknowledges research support from the Markle Foundation.

Events Leading to Divestiture

AT&T emerged from World War II with a monopoly of domestic sales of telephone equipment and long-distance service. Its monopoly of local exchange service was not as complete, but only because for the most part more than a thousand small, mainly rural independent telephone companies and cooperatives served regions where the telephone business was least profitable.

Historical Developments: Technology

The monopoly structure of the telephone industry at the end of World War II was believed to be driven by its technology. In 1945, the telephone system consisted of copper wires and electromechanical switches. The local telephone network connected each telephone in service to a local exchange switch by means of a dedicated pair of copper wires. These pairs were bundled into large cables. Installing cables involved considerable fixed costs that were independent of the cable's capacity, so that unit costs of capacity tended to decline with the total capacity of the cable. Long-distance service was provided by still more copper cables, which linked local switches to a hierarchy of long-distance switches. The trunks, too, displayed scale economies for the same reasons. The switches, however—at least, those beyond a relatively small minimum efficient size—did not exhibit scale economies.

Analytically, of course, scale economies in cables are insufficient proof of natural monopoly, for a telephone company is a complex organization providing a variety of services over its network, for which diseconomies of scale or integration might well be present. Moreover, the monopolization of the telephone business had not in any obvious way arisen because of the push of natural monopoly efficiencies. At the turn of the century, competition in local exchange systems was common in larger cities. The cornerstone of the Bell System monopoly rested on patents for the only workable long-distance technology, combined with a practice of denying interconnection with the Bell long-distance system to non-Bell local exchange companies. Nevertheless, by 1945, federal and state policy was firmly grounded in the premise that the system was a natural monopoly.

The period between World War II and 1974, the year the Department of Justice filed its lawsuit, was marked by four techno-logical developments that significantly affected regulatory policy. The first, growing out of wartime research, was the use of microwave relays in place of copper wires for long-distance communication. The second, also influenced largely by the government through the 1960s and early 1970s, was the development of communications satellites for domestic transmission of telephone traffic. The third, at about the same time, was the replacement of electromechanical switching machines with electronic switches that were essentially computers. Finally—also in the early 1970s—came the development of cellular mobile telephone technology.[1] While not the only technologically significant developments growing out of the telephone industry (Bell Labs having, for example, invented the transistor during this period), these four developments were important because of their impact on the structure, or potential structure, of the telephone industry.

The significance of both microwave and satellite technologies was not so much that they lowered in any obvious or compelling way the minimum efficient scale of firms producing telecommunications service. Rather, their importance lay in the fact that both technologies required a federal license to use the electromagnetic spectrum. Issues governing efficient use of the radio spectrum forced the government to consider policy questions it had not seriously considered since the establishment of the second Bell monopoly by Theodore Vail. In the late 1950s, numerous companies applied for microwave licenses to build private systems for their internal needs. This development raised the policy question of whether the FCC should permit competition in long-distance service.

The FCC initially decided to permit limited entry through vertical integration by large customers and then went on, a decade later, to permit entry by so-called specialized third-party microwave carriers providing private line services to large customers. The fundamental premise behind these decisions was that the telecommunications industry was divisible into two discrete parts: that which was a natural monopoly (ordinary switched local and long-distance telephone service) and that which was not, and thus ought to be competitive (private lines, the distribution of broadcast signals by networks to affiliates, "enhanced" computer services, and so on). By the early 1970s, the FCC had reached the same conclusions about the domestic communication satellite business, in which it not only permitted open entry but initially forbade AT&T to use the technology for any purpose other than providing ordinary switched message telephone service (MTS).

Electronic switching technology was important for two reasons. First, it caused a convergence of telecommunications and computer/electronic equipment manufacturing, bringing potential competitors to the telephone equipment business. Second, computer-based switches vastly increased the range and value of services that could be offered by telephone companies.

Finally, the significance of cellular radio technology, apparent only recently, lies in the potential of systems that use the electromagnetic spectrum to serve as an economical alternative to copper wires in the provision of local exchange service, permitting local competition for the first time since the early years of the twentieth century. Cellular mobile radio systems now in use are technically inappropriate and may have been allocated insufficient amounts of the electromagnetic spectrum, in order to displace local copper wire systems serving fixed locations. But the general approach of using radio telephony, optimized to serve fixed stations, potentially could challenge the principal remaining rationale for telephone regulation, namely, the monopoly status of local distribution systems for local telephone service. Whether the challenge from cellular radio technology will become real depends not only on policy decisions but also on whether other new technologies, such as "pair gain"[2] or fiber optics, prove less expensive for basic access and display significant economies of scale so that local access remains a natural monopoly.

Historical Developments: Regulation

The FCC was the focus of the policy debates on telephone industry structure throughout the postwar period. Indeed, until the announcement by Assistant Attorney General William Baxter and AT&T Chairman Charles Brown of the historic settlement of *United States v. AT&T*, few observers believed that either the antitrust case or concurrent legislative efforts would result in significant or timely change.

What the FCC gradually and timidly accomplished between

1945 and 1980, with prodding from the courts and Congress, was to permit entry into the long-distance service and telephone equipment businesses. Entry attempts began almost immediately after the war. With respect to equipment, attempts at entry began with the *Recording Device* cases in 1947, followed by *Hush-a-Phone* in 1956, and culminating in *Carterfone* in 1968. Throughout, AT&T took the position that to permit customers to attach "foreign" (non-Bell) equipment to the telephone network would result in degraded service and danger to the safety of telephone employees. It is difficult today to understand how these arguments could have been taken seriously for more than twenty years. At the time, however, the burden was on those favoring competition to prove that harm would *not* occur. Proving these negatives was difficult before there was experience with competition. Throughout the debates on telecommunications competition, placing the burden of proof on would-be entrants worked to the advantage of AT&T and may account for the glacial pace of telecommunications regulatory reform.

In long-distance service the FCC initially permitted private microwave systems and then, in the late 1960s, permitted entry by specialized carriers, using as its excuse the new entrants' desire to provide types and qualities of private line service that AT&T did not choose to offer. The new entrants quickly demonstrated a disrespect for this rationale by providing ordinary long-distance telephone service.

AT&T's major policy defense of its long-distance monopoly was the claim that the monopoly was a natural one, justified by economies of scale. Entry, according to AT&T, was attractive to outsiders only because of AT&T's system of internal subsidies, which kept long-distance prices high in order to maintain lower prices for local service. It was not clear at the time whether the subsidy really ran in the direction claimed by AT&T.[3] More to the point, there was never any demonstration that the alleged subsidy was socially or economically desirable. Inefficient pricing resulting from the alleged subsidy was defended on two grounds. First, it was argued that all subscribers benefit if another subscriber joins the network. This notion was supposed to justify the quest for "universal service" because of the "externality value" of subscription. But residential demand for basic telephone service is almost perfectly price inelastic, and nearly everyone subscribes, factors greatly vitiating AT&T's argument. If guaranteeing that the last few percent of households actually subscribe is valid social policy, a targeted subsidy would be far cheaper and less inefficient than a cross-subsidy. The second defense of the alleged subsidy was based on the presence of beneficial income distribution effects, but the distributional effects of a subsidy running from long-distance to local service, while not entirely clear, do not appear to provide significant net benefits to the groups that they are intended to help.[4]

Throughout the debates on whether entry should be permitted in the sale of telephone equipment and long-distance service to consumers, advocates of competition assumed that local exchange service was, if not a natural monopoly, at least a permanent franchised monopoly that would be perpetuated by state regulation. Indeed, because AT&T controlled access to this bottleneck the corporation was viewed as being able to monopolize long-distance and equipment markets.[5] Local exchange customers who bought equipment elsewhere would have their phone service cut off; they had no alternative supplier. Similarly, customers who wanted to deal with a competing long-distance supplier would be either denied local connections to that supplier or charged a high price for such connections. Alternative connections that did not use the local exchange were either unavailable or much more expensive.

By the early 1970s, the FCC had authorized entry into virtually all federally regulated services except ordinary switched MTS, an area in which competition was permitted a few years later. New entrants complained vigorously that AT&T was inhibiting competition by engaging in two types of dirty tricks. In long-distance, competitors complained that AT&T was engaging in pricing practices designed to drive them out of the market. In particular, AT&T was alleged to use its regulated monopoly in long-distance MTS to cross-subsidize the competitive specialized businesses, at that time mainly private line services to large customers. In addition, AT&T's competitors alleged that the local Bell operating companies (BOCs) were delaying or denying local exchange interconnection not only to customers using competing equipment but also to competing long-distance carriers seeking to be connected to their local customers. In 1971, 1972, and 1973, these complaints piled up at the FCC and the Department of Justice.

AT&T's motives for engaging in anticompetitive behavior

against the new entrants were derived not only from its managers' conviction that a telephone monopoly was socially desirable but also from the incentive structure created by regulation itself. To begin with, the overall profit of the company was limited to the estimated cost of capital times the depreciated book value of its "used and useful" installed base of capital equipment. In addition, most of the rate base of the company was for local service, which was regulated by the states. With few exceptions, regulators reviewing prices ignored service-specific costs and did not, until late in the game, attempt even to estimate them. Most extreme was the practice in nearly all the states of "residual pricing." That is, regulators would first extract as much revenue as economics or politics allowed from most telecommunications services and then collect the remaining "residual" of costs by setting prices for the most politically visible items: (a) installation and basic monthly service for residences and (b) pay telephone calls.

The FCC did not adopt residual pricing explicitly; indeed, for all practical purposes the FCC had no policy regarding the structure of telephone service prices. For more than thirty years, the FCC undertook no formal review of AT&T's price structure. Then, in response to the complaints of competitive entrants, the commission found itself unable to act swiftly and decisively, because it lacked not only sufficiently detailed and reliable cost data but even a policy.

In one respect, telecommunications regulation was a great success: monthly residential phone prices in real terms declined steadily as more telecommunications services were developed, and penetration of the telephone system grew to almost 100 percent. Moreover, as noted above, regulation did restrain monopoly pricing, for all estimates of the elasticity of demand for residential service at prevailing prices are remarkably small (about -0.01 to -0.10), indicating that there was scope for enormous increases in the revenues of telephone companies from increases in basic monthly charges.

In this environment, competition posed little threat to even an inefficient monopolist so long as the regulatory system remained unchanged. Preserving a market against successful entry allowed the regulated firm to continue to make used and useful investments to serve it. These investments would then be allowed into the rate base for calculating profits. Moreover, because individual prices were not set on the basis of costs, the monopolist could match or beat any entrant's price (even at a price below the monopolist's costs in that market) and then make a compensating adjustment in prices elsewhere, cross-subsidizing the competitive market, the so-called "other Averch-Johnson effect."[6]

The main difficulty with AT&T's strategy for dealing with competitors was that, in a regime of residual pricing, the compensating price changes had to be made in the tariffs most cherished by state regulators, thereby threatening the political accommodation between the utility and its regulators. Although federal law technically bars regulators from expropriating a firm's capital, there remain ways that regulators can resist passing on to residential rate payers all the costs incurred through losses in competitive markets.

For these reasons, taking actions to disadvantage competitors was even more attractive to AT&T and its subsidiary BOCs than cross-subsidization was. The entrants had to connect to the local monopoly to provide service. If these connections could be denied, cross-subsidization would be unnecessary. Or, if regulators were to force interconnection, the local telephone company might be able to take actions that raised the costs of entrants relative to the competitive offerings of the local monopolist. The local telephone company could impose excessive technical compatibility requirements, technically inferior conditions of interconnection, or discriminatory prices to competitors for their use of the local monopoly.

The FCC's gradual and timid experiments with freer entry had been encouraged by the Department of Justice, which in the mid-1960s instituted a competition advocacy program aimed at reform of federal regulatory agencies, including the FCC.[7] Of course, the Department of Justice had brought an antitrust case in 1949, *United States v. Western Electric,* seeking divestiture of AT&T's equipment-making subsidiary. That case was settled in 1956, when President Eisenhower ordered the Department of Justice to leave the telephone company alone. The settlement forced AT&T to license its patents and restrict itself to its then-monopoly businesses. By the 1970s, the decree was widely viewed as anticompetitive, because it kept AT&T out of the computer business. Also, using the decree as an excuse (although its terms did not forbid it), AT&T declined to attempt to compete vigorously in the overseas sale of telephone equipment.

With this background, it is easy to understand the motives of the DOJ in bringing the antitrust lawsuit in 1974. Some of the incentive and much of the ability to engage in anticompetitive behavior would be eliminated if the local exchange monopolies were divested from the competitive, or potentially competitive, segments of AT&T that provided equipment and long-distance service. And some of what the Department of Justice perceived as competitive problems in the computer industry would be ameliorated, if not eliminated, by freeing AT&T to compete with IBM. Thus, the antitrust lawsuit was brought to achieve these goals.

In seeking divestiture along the lines it did, the Department of Justice implicitly assumed that there were no significant economies of vertical integration within AT&T, although no evidence on this point existed. The department did not, however, seek to dissolve either Long Lines or Western Electric into smaller units, because it was unwilling to make a similar assumption about economies of scale within these businesses. The government sought to make the world safer from AT&T's predatory tendencies, not directly to create a structurally unconcentrated marketplace. The supposition was that if new entrants did not succeed in taking market share from AT&T through superior efficiency, they would at least serve as potential competitors, forcing AT&T to rationalize its pricing policies.

One did not have to hold unrealistic views on the effectiveness of potential competition to believe it more effective than FCC regulation. Indeed, the Department of Justice was willing to seek structural relief despite the risks and delays it would face partly because the alternative remedy—regulation—was demonstrably ineffective in dealing with AT&T's inefficient and anticompetitive pricing behavior. The FCC could not determine AT&T's costs, nor could it settle on a sensible cost-based method for pricing. One set of AT&T prices, the Telpak tariff, went through nearly two decades of hearings without a final determination of its lawfulness. It was apparent that even with a fully informed regulatory policy and the best will possible, the FCC could not cope successfully within available administrative procedures with AT&T's control of the information necessary to regulate prices effectively.

The Relief Sought by the Government

At trial, the government sought divestiture of local exchange service from the rest of AT&T. In the early stages of the litigation, the objective was divestiture of the twenty-three BOCs. Later, when it became apparent to the government that the BOCs controlled substantial long-distance assets, the government proposed a so-called functional divestiture that focused on the physical division between interexchange and intraexchange services.

The government also sought an injunction preventing reintegration by the operating companies into competitive businesses. Because regulated monopolies can effectively evade regulatory constraints by engaging in below-cost pricing and discriminatory acts in otherwise competitive markets, Assistant Attorney General Baxter took the view that the operating companies should be barred entirely from any but regulated monopoly businesses. This policy echoed the old 1956 *Western Electric* decree. Such a policy is difficult to implement because of the necessity for distinguishing between BOC participation in a competitive business and BOC adoption of a new production technology or a product innovation in its existing business. Presumably interfering with the latter is undesirable. Moreover, successful cross-subsidization typically involves juggling the books to fool the regulators. While such juggling has been easy enough for telephone companies operating telephone-related businesses within their own territory and with common employees and equipment, it would presumably be much more difficult (as would discriminatory practices) outside that territory in cases in which the production facilities and employees are not used in common with regulated services.[8] The settlement thus embodied an extreme, purist approach to the cross-subsidy problem.

The government sought to have a reasonable number of operating companies owned separately in order to avoid monopsony problems in the purchase of telephone equipment. This form of divestiture also created several powerful potential competitors for each operating company in areas in which local exchange carriers were likely to compete, such as directories. The seven regional companies also might eventually be effective competitors against the remaining parts of AT&T when they are permitted to enter such lines of business. Moreover, by making the seven regional companies independent, the government created an opportunity for state regulators to make greater and more effective use of yardstick competition. In addition, the government requested injunctions requiring the BOCs to provide equal access to all long-distance companies; the purpose of this objective was to put the other common carriers on an equal footing with AT&T in competing for long-distance business.

Expectations

The relief toward which the government labored for seven years in its lawsuit was sharply distinguished from earlier cases under Section 2 of the Sherman Act. The traditional government-originated monopolization case had been brought to impose structural competitiveness upon an industry. In *Standard Oil* and *American Tobacco,* monopolists were broken into three or four competing firms. In *Alcoa,* government-owned plants were sold under court supervision to competing firms, one newly created. But in *United States v. AT&T,* the relief did not seek to create any new competing firms, because the government could not simply dismiss the possibility that economies of scale were present. The government certainly could not have proved, even to its own satisfaction, that long-distance service was not a natural monopoly.

Nevertheless, it would certainly have been quixotic to spend the time and resources that were spent on *United States v. AT&T* merely in the hope that potential competition would improve upon regulation in inducing efficient performance from AT&T.[9] The staffs (turnover was substantial) prosecuting the case believed that the decree they sought would eventually lead to a less concentrated, more competitive structure in long-distance service and in the markets for telephone equipment.

The Consent Decree

The private agreement announced by the parties at the end of 1981 gave the government literally all that it had asked for. U.S. District Judge Harold Greene, however, declined to accept the settlement without public scrutiny and analysis. After receiving comments from the public, the jurist modified the decree in at least two significant respects. First, AT&T was barred for seven years from offering any electronic publishing services, and the Yellow Pages business that the settlement had left with AT&T was handed over instead to the operating companies. Second, the operating companies were permitted to sell (but not manufacture) telephone equipment. Having made these and several other less important changes in the settlement, Judge Greene entered his judgment as a modification of the old 1956 *Western Electric* decree, so that the earlier restrictions on AT&T were repealed. The decree thus became known as the Modified Final Judgment (MFJ) in *United States v. Western Electric.*

Criticism of the MFJ

The decree, although it gave the government most of what it asked for, is far from perfect. Moreover, several of the modifications of the settlement introduced by Judge Greene were unsound.

The major imperfections of the decree from the point of view of what the government had requested are related to BOC participation in other lines of business and the implementation of the functional divestiture of Long Lines from the operating companies.

With respect to permissible "lines of business," the government simply did not propose a workable method of identifying those businesses which the BOCs could enter. The purist view of Assistant Attorney General Baxter could not be defended against the charge that the government had been overly restrictive in proscribing, for example, economic activity by the BOCs outside their territories in areas unrelated to telecommunications. Moreover, popular opinion at the time of the announcement of the settlement—inexplicably influential with Judge Greene as well as with state and federal regulators—was concerned with the possibility that the operating companies would flounder in the postdivestiture era. Finally, the Department of Justice did not correctly anticipate the effect of its disposition of AT&T's licenses, already granted by the FCC, to provide cellular radio telephone service throughout the nation. These licenses were assigned to the operating companies. Had the licenses remained with AT&T, cellular radio might have been the basis for the development and promotion of radio-based local distribution systems as a competitive alternative to wireline local exchange service.[10] Today the BOCs not only control the cellular licenses allocated to wireline carriers in their service territories but also, with the consent of the Department of Justice, have engaged in a pattern of acquisitions of nonwireline licenses outside their territories.[11] Collectively, the BOCs now stand to dominate the most promising competing technology for local exchange service. They also sponsor their R&D through a joint venture, Bell Communications Research, so that no BOC need be concerned that another BOC will spring forth with a competitively advantageous technology. The BOCs are most likely to develop cellular technology as a complement to, rather than as a substitute for, their existing services.

The second major imperfection in the MFJ lies in its implementation of the functional divestiture of local services from long-haul services. The department's relief proposal at trial contemplated divestiture of *local* exchange from the rest of the network. Local meant the lowest level of the switching hierarchy, the number 5 switching offices. For technical reasons that Assistant Attorney General Baxter was willing to defend,[12] and without objection from the competing carriers, the definition of local exchange was expanded to include all territory served by a number 4 switching office, with some subdivisions to take account of state regulatory jurisdictions and some aggregations to make a Consolidated Metropolitan Statistical Area (CMSA) the smallest possible local exchange territory. As a result, the concept of divestiture of the local exchange metamorphosed into divestiture of Local Access and Transport Areas (LATAs) that were of significant size (several are entire states), containing substantial toll markets. A continuing widespread BOC monopoly of intraLATA toll was an inevitable feature of the MFJ because state regulatory officials tend to be skeptical of competition. In states where intraLATA monopoly is most rigorously enforced, this feature of the MFJ also served to disadvantage AT&T's competitors in the long-distance market, as discussed below.

One imperfection in the decree introduced by Judge Greene is the anticompetitive restriction on AT&T's provision of electronic publishing services. More significant, however, was Judge Greene's decision to permit the operating companies to sell customer equipment and retain Yellow Pages advertising. This decision was probably an effort to cater to the transitory public perception that the BOCs should be pitied. In letting the BOCs sell equipment and retain Yellow Pages, the judge permitted the BOCs to participate in competitive businesses closely related to their local monopoly, using common resources. In neither case was there an argument that failure to permit BOC participation would restrict supply significantly or limit the BOCs' ability to adopt new technology. From the perspective of the economic theory of the case, these modifications were unsound. Whether they will have a serious adverse effect on the public is, however, still unclear.

Imposing these changes in the settlement agreement had another price, one that Judge Greene may have come to regret. Each time the operating companies wanted to enter a new line of business, they were required to obtain a waiver from the judge. But the abandonment of Assistant Attorney General Baxter's purist rule, and the failure to replace it with a coherent or consistent policy for deciding which lines of business were permissible, meant that "line of business" waivers became a decisional quagmire.

Events Since Divestiture

Divestiture itself did not create a competitive market in any part of the telephone industry. It did not, as in *Alcoa*, instantaneously create new competitors, nor did it, as in *Standard Oil*, divide any part of AT&T along horizontal lines. Instead, it focused on eliminating what the government regarded as the primary institutional barrier to effective competitive entry: the vertical connection between franchised monopolies in local service and the other parts of AT&T in which the continued dominance of AT&T could plausibly be ascribed to the perverse incentive structure of regulation. In this sense, AT&T divestiture is a grand experiment in the "new learning" in the economics of industry and antitrust, one that downplays the role of market structure (e.g., measures of market shares of leading firms), concentrating instead on the possibility of entry in response to inefficient performance by incumbent firms.

At the time of divestiture, AT&T possessed extremely high market shares in almost all sectors of domestic telecommunications. Nevertheless, other firms were poised to compete in all important lines of business except basic local exchange service for residences and small businesses. But it was unclear whether the entrants could match the efficiency of AT&T, a firm with decades of experience as the dominant firm and, more formidably, with gigantic Bell Labs, the jewel of industrial research, throwing its weight against the technological frontiers of the industry. It was also unclear whether state regulators would resist competition actively or only passively; that they would resist was a foregone conclusion. As we shall argue, the advocates of competition were proved right about the ability of entrants to challenge AT&T, for in numerous markets AT&T proved to be a vulnerable monopolist indeed, owing to high costs and poor management practices. Competition has faced tougher sledding, however, in the long-distance business. The rise of competition has been in the face of persistent and organized resistance by state regulators (with a few notable exceptions). Nevertheless, we believe that the Department of Justice's grand experiment has been largely successful—remarkably so, considering the stance of state regulators and the defects in the implementation strategy of divestiture.

Long Distance Service

As noted above, divestiture made something of a mess of the long-distance service business by creating LATAs that, in most cases, are so large that about 25 percent of all long-distance revenues are generated from toll calls completed within LATAs.[13] The Department of Justice had defensible technical reasons for constructing LATAs as it did. Unfortunately, it simultaneously created a new de facto political boundary, within which state regulators would have virtual autonomy in deciding the role of competition. Although LATA boundaries were not intended to demarcate the boundary between monopoly and competition,[14] such has been the result in nearly all cases.

State regulators can block intraLATA competition in any of three ways. First, they can simply deny facilities-based competitors the right to make intraLATA connections, subjecting the competing interexchange carriers to penalties or threatening to remove their licenses to provide intrastate, interLATA service if they permit intraLATA calls to be made over their networks. Second, state regulators can adopt "block or pay" policies, which allow an interexchange carrier to permit its customers to make intraLATA calls but require it to pay the local exchange carrier either (a) all its revenues or (b) even all the revenues that the latter would receive had the call been placed over the local exchange carrier's network. (The latter policy is more Draconian, for local exchange carriers usually charge more for toll calls than interexchange carriers do.) Third, the state regulators can permit competition but subject the interexchange carrier to high "carrier access charges" for intraLATA calls. To offer intra-LATA, interexchange service, an interexchange carrier will have to pay the local exchange carrier a fee—sometimes per local exchange carrier customer but more commonly per minute of intraLATA call—that is so high that the interexchange carrier must either lose money on intraLATA service or set prices substantially above its costs and the local exchange carrier's prices.

Most states do not beat around the bush; they simply outlaw intraLATA competition. As of January 1987, only fourteen of the fifty-one jurisdictions (including the District of Columbia) allowed facilities-based (as opposed to resale) intraLATA competition. Of these, three jurisdictions effectively prohibit competition by imposing a "block or pay" rule,[15] and several others restrict the extent of permissible competition or simply have failed to license any competitors.[16] In some states, the ban is rather ineffective, for interexchange carriers can let their customers make intraLATA calls as long as the interexchange carriers keep secret that such calling is possible. They cannot advertise that such service is available or publish their prices for such calls. Moreover, they are not provided with equal access for intraLATA service, and so they cannot offer comparable service quality.

The anticompetitive policies of most state regulators are significant because intraLATA toll is both important and overpriced. As dictated by "residual pricing" policy, intraLATA toll is viewed by state regulators as a source of cash for holding down basic monthly service charges. Of the $60 billion in revenues collected for all toll calls in 1986, $15 billion was collected by local exchange carriers—the seven BOCs,[17] along with independents such as GTE.[18] The interexchange carriers, after paying access charges to the local exchange carriers, collected net revenues of $25 billion, while placing approximately three times as many calls over greater distances.[19] Thus, it is reasonable to conclude that nearly half the revenues from local exchange carrier toll service are due to prices in excess of costs, at least the costs that would be faced by an interexchange carrier.

The anticompetitive policies of most states also matter because they inhibit development of robust competition among LATAs. Preventing intraLATA calls is a serious problem for all interexchange carriers except AT&T. Divestiture left AT&T with boundaries for service territories that correspond to LATAs, since it was AT&T's old network that was divested. Thus, if AT&T simply did nothing, it would handle almost exclusively only legal toll calls. The other interexchange carriers, however, are not so fortunate. Their networks were designed according to their own prior perception of toll markets and did not neatly match the LATA boundaries. To obey state anticompetition rules, therefore, not only requires AT&T's competitors to incur costs to reengineer and/or reprogram their networks but also leaves interexchange carrier network design maladapted to the new set of permissible and impermissible markets. State regulators have inadvertently erected a barrier to competition against AT&T's competitors while purposely trying to prevent all interexchange carriers (including AT&T) from effectively entering intraLATA markets.

The divestiture plan itself created still another problem for AT&T's competitors. Divestiture imposed upon the BOCs the obligation to provide equal access to all interexchange carriers in most of their service territories within a few years. The policy was to require equal access in all electronic switches (where it was easiest and cheapest), as well as in other switches that served a large number of customers or that otherwise were not so costly to reconfigure that to require equal access would be irrational. No other requirements were placed on the sequence in which switches would be changed, nor was a clear definition offered as to what constituted a cost that was unreasonable to bear for the purpose of providing equal access.

The BOCs, acting sensibly, went about the task of providing equal access in a manner that more or less minimized their costs of compliance with the equal access requirement. This meant making inexpensive changes first and, where possible, combining the provision of equal access with normal replacement investment or major maintenance in switching centers. It also meant engaging AT&T's competition in hard bargaining over which switches would be exempt from early compliance—bargaining that periodically ended back before Judge Greene.

The result of this policy has been two factors that inhibit competition in long distance. First, within a given market area, different switches were converted to equal access at different times, sometimes years apart. This patchwork conversion of switches raised the costs (per prospective customer) of mass media advertising by AT&T's competitors,[20] for the other common carriers could not target their message solely to customers who were in the magic period of commitment.

The second problem with the plan was that it prevented long-distance carriers other than AT&T from fully rationalizing their own construction plans for national networks. For about a decade after divestiture, the other common carriers will not be able to offer national equal access service, for a portion of the local network will still not have been converted. By the end of 1986, about three-fourths of the BOCs' telephone customers were connected to a switch with equal access;[21] the worst performance was by NYNEX, which was only a little above one-half converted to equal access (see Table 5-1).

Dissatisfied with the speed with which equal access was being provided, MCI complained to Judge Greene in 1986. In examining the issue, the Department of Justice concluded that in several instances the delays in conversion were not adequately justified. This conclusion applied to 21 central offices of Southwestern Bell, 77 offices of BellSouth, about 300 offices of NYNEX, and at least 100 and maybe 200 offices of US West.[22] In the first two cases, the magnitude of noncompliance, even with worst-case assumptions, is small, representing 2 to 3 percent of total central offices.[23] For NYNEX, the number of offices in dispute represents nearly 25 percent of all NYNEX offices. For US West, between 10 and 15 percent of its offices are in dispute. Thus, five of the seven BOCs have apparently made a reasonable attempt to comply with the divestiture requirement, while two raise serious questions about foot-dragging.

More significant than keeping score on compliance is the fundamental plan embodied in the divestiture requirement. Two aspects of the policy are of dubious validity. The more obvious is the slow pace of conversion, combined with its hit-or-miss geographic pattern, that is accepted by the Department of Justice.

Table 5-1

Total End Offices and Lines Providing Equal Access as of September 1, 1986

Company	End Offices		Lines	
	Number	%	Number (000)	%
AMERITECH				
Illinois	164	53.2	3,834.2	79.6
Indiana	50	30.7	1,062.5	74.0
Michigan	126	33.2	2,725.7	71.3
Ohio	123	47.5	2,222.8	75.0
Wisconsin	74	57.8	1,357.3	87.4
TOTAL	537	43.4	11,202.5	76.7
BELL ATLANTIC				
Bell of Pa.	262	65.2	4,219.0	89.5
C&P—Md.	74	31.8	1,312.0	53.9
C&P—Va.	153	70.8	1,938.0	87.3
C&P—W. Va.	28	18.7	291.0	45.2
C&P—D.C.	22	75.9	701.0	89.3
Diamond State	33	100.0	353.0	99.7
N.J. Bell	118	55.9	3,448.0	79.4
TOTAL	690	54.2	12,262.0	79.2
BELLSOUTH				
South Central Bell	253	26.4	4,039.6	62.1
Southwestern Bell	323	49.5	6,402.1	76.9
TOTAL	576	38.0	10,441.7	69.5
NYNEX				
New York Telephone	254	36.6	4,931.3	54.2
New England Telephone	156	23.6	2,821.9	58.0
TOTAL	410	30.2	7,753.2	55.5
PACIFIC TELESIS				
Pacific Bell	306	42	8,390.2	72.0
Nevada Bell	10	23.8	132.0	72.4
TOTAL	316	32.9	8,522.2	72.2
SOUTHWESTERN BELL				
Southwestern Bell Telephone	411	31.6	8,202.0	74.2
US WEST				
Mountain Bell	216	26.8	3,893.0	74.5
Northwestern Bell	114	17.5	2,113.3	61.8
Pacific Northwest Bell	103	39.8	2,123.7	82.3
TOTAL	433	28.0	8,130.0	72.9

Source: "Report of the United States to the Court Concerning the Status of Equal Access," (U.S. Department of Justice, November 21, 1986), p. 5.

Less obvious is that the costs of conversion are being imposed on literally everyone *except* the firm that derives most of the benefit from the pace of the divestiture plan—AT&T.[24] The conversion requirements, at least as interpreted in 1986 by the Department of Justice, prolong the period in which AT&T enjoys a competitive advantage. Meanwhile, the costs of conversion are paid by the BOCs, whose only benefit from conversion is that the ability of the other common carriers to compete for intraLATA toll business is retarded somewhat. That benefit, however, is relatively small because most state policies prevent this competition.[25] In addition, the policy imposes costs on the other common carriers by distorting the pattern of demand on their own networks and by delaying the date when they can compete on equal footing with AT&T. Ameliorating the anticompetitive effects of the pattern of compliance with the MFJ's equal access requirements is the fact that many of the exchanges to which the other common carriers do not yet have equal access are, for reasons of marketing or location, less desirable to the other common carriers than most of the exchanges that have been converted.

The treatment of long distance in the divestiture agreement makes for an inevitably long period of transition to competition. Nearly twenty years after the FCC first opened the door, and five years after the announcement of a settlement agreement, the long-distance market has not yet provided a fully fair test of the viability of competition. About a quarter of the market is monopolized by the BOCs, another 25 percent lacks equal access, and the remaining half is affected both by the way conversion has been carried out and by the presence of competitive disadvantages in the other half. Realistically, it will be several more years before the full effects of the experiment with long-distance competition can be confidently and comprehensively assessed.

Nevertheless, as of mid-1987 the conventional view held that competition in long distance has not been very successful. In 1986, Wall Street turned bearish on AT&T's competitors.[26] Peter Huber opined in his 1987 report that the other common carriers' financial performance was poor and that "many if not all" other common carriers stay in business at the sufferance of AT&T and the FCC.[27] The basis for these conclusions lies in two facts: (a) the continued high market share of AT&T and (b) the low earnings of the new competing carriers. In neither case, however, do the facts reveal a serious threat to the durability and effectiveness of competition compared with regulated monopoly. From the perspective of the success of divestiture, the appropriate standard is not whether AT&T has a large market share or whether AT&T is more profitable than the new competing carriers. Instead, it is whether the other common carriers are financially viable in the long term and whether they more effectively force AT&T to set efficient prices than the FCC did when AT&T enjoyed a monopoly. On both counts, the Department of Justice experiment shows every sign of being a success.

The basic market share data are displayed in Table 5-2. There we show three independent sources for estimates of market shares from 1985 to 1990. These estimates diverge rapidly, but by any measure AT&T will be a dominant firm well into the 1990s. The key to these analyses, however, is the basis for the divergence. The firm estimating the highest AT&T share gives as its explanation that it expects AT&T "to show only moderate revenue growth through 1990 as it cuts rates."[28] The others are simply less sanguine about AT&T's ability to adjust effectively

Table 5-2
Share Estimates: InterLATA Toll

	1984	1985	1986	1987	1988	1989	1990	1991
GOLDMAN, SACHS								
AT&T	91.0	87.2	83.0	78.1	74.7	71.4	68.7	66.2
MCI	6.4	7.0	8.3	9.9	11.4	12.9	14.0	15.2
US Sprint	1.3	3.5	4.4	5.9	7.2	8.2	9.0	9.6
OTHERS	1.3	2.3	4.4	6.1	6.7	7.5	8.3	9.0
DEAN WITTER								
AT&T	—	86.2	82.5	79.0	76.7	74.8	73.3	—
MCI	—	6.8	8.6	10.5	11.5	12.5	13.4	—
US Sprint	—	3.7	5.1	6.3	7.4	8.1	8.6	—
OTHERS	—	3.4	3.9	4.2	4.5	4.7	4.7	—
BAIRD								
AT&T	—	87.0	83.0	81.0	—	—	80.0	—
MCI	—	5.0	8.0	9.5	—	—	10.0	—
US Sprint	—	3.0	4.5	5.5	—	—	7.0	—
OTHERS	—	5.0	4.5	4.0	—	—	2.0	—

Sources: Daniel F. Zinsser, *Telecommunications Quarterly,* June 30, 1986, Goldman, Sachs Research; "MCI Communications Corporation: Boy Did We Get a Wrong Number," Research Note 1342, August 20, 1986, Dean Witter Reynolds, Inc.; and "MCI Communications Corp.," *Technology Perspective: The Telcom Report,* August 26, 1986, Robert W. Baird & Co.

Table 5-3

Revenues and Profits of Interexchange Carriers (1986: First Half)

	Revenues (millions of dollars)	Net Profits (millions of dollars)
AT&T Communications	$18,570	$742
MCI	1,762	36
GTE Sprint	826	−102
US Tel	212	−40
ALC Communications	238	−6
RCI	18	−7
TEL/MAN	21	1

Source: "MCI Communications Corporation: Boy, Did We Get a Wrong Number," Research Note 1342, August 20, 1986, Dean Witter Reynolds, Inc.

to competition from other carriers. The outcome turns on *how* AT&T responds to the threat of the competitors, not *whether* it is subjected to competitive pressure. AT&T faces a choice of strategies. At one extreme, it may preserve market share and growth by charging competitive prices, foreclosing the growth of the other carriers as the institutional and technical barriers to competition erode. At the other extreme, AT&T can price to earn transitory excess profits from (a) a diminishing base of "captive" customers over whom AT&T enjoys a temporary advantage and (b) those who will be the slowest to respond to more attractive competitive options.[29] Regardless of the outcome, competition eventually will have worked by lowering prices and improving the efficiency of the long-distance market.

Of course, for this rosy story to work out, some competitors must survive into the 1990s, either to keep AT&T's prices low or to take away market share if AT&T prices monopolistically or operates inefficiently. One basis for widespread pessimism on this score is the current lack of profitability of the competing carriers, exemplified by the data in Table 5-3. To obtain some perspective on these data, Table 5-4 shows the details of the actual and estimated performance of MCI, taken from one of the pessimistic analyses of the prospects for the new carriers. This report shows almost constant net profits for MCI throughout the period 1982–1988 (measured by net income) and declining earnings per share, punctuated by a large loss in 1986. But notice three other items: gross interest expense, depreciation, and asset write-offs in 1986. The first measures the amount of capital

Table 5-4

MCI Communications Projected Income Statement

($ In Thousands, Except Earnings Per Share)

	Actual					Estimated	
	1982	1983	1984	1985	1986	1987	1988
Total revenues	$906,596	$1,521,460	$1,959,291	$2,542,200	$3,592,000	$4,200,000	$4,900,000
Less:							
Local interconnection	142,972	262,012	479,658	873,900	1,636,000	2,079,000	2,450,000
Leased facilities	104,438	273,663	343,257	280,500	267,000	189,000	220,500
Internal expenses	307,951	487,254	696,324	834,600	1,097,000	1,100,000	1,200,000
Interest expense	69,141	134,277	188,545	201,100	187,000	215,000	225,000
Depreciation	89,164	158,959	264,573	347,200	451,000	500,000	585,000
Restructuring							
Asset write-down	—	—	—	—	453,000	—	—
Other	—	—	—	—	132,000	—	—
Plus:							
Interest income	21,602	76,708	114,644	85,300	63,000	30,000	20,000
Other income (expense), net	(729)	365	(51,044)	77,600	71,000	0	0
Pretax income	$213,803	$282,368	$50,534	$167,800	(497,000)	$147,000	$239,500
Income tax:							
Current	22,246	14,254	438	700	8,000	24,400	47,900
Deferred	40,142	65,202	(9,107)	27,600	(57,000)	5,000	23,950
Effective tax rate	29.2%	28.1%	-17.2%	16.9%	*	20.0%	30.0%
Net income	$151,415	$202,912	$59,203	$139,500	(448,000)	$117,600	$167,650
Earnings per share							
Primary	$0.78	$0.89	$0.25	$0.59	(1.63)	$0.44	$0.62
Fully diluted	$0.73	$0.88	$0.25	$0.59	(1.63)	$0.44	$0.62

*Negative income and taxation

Source: "MCI: The Sword of Damocles; Opinion-Sell," *Equity Research Weekly Research Notes*, December 11, 1986, First Boston Corp., p. 3, except for 1986, which is from *MCI Communications Corporation Annual Report 1986*.

investment accounted for by borrowing, while the latter two measure (imperfectly) capital consumption and represent additional cash flow. The sum of depreciation and write-offs represents the accumulation of additional cash to investors beyond measured net income. This accumulation plus net interest represents additional returns to investors that are not included in net income. Thus, the First Boston figures show MCI's cash flow growing from $241 million in 1982 to $487 million in 1985 (actual) and to $753 million in 1988 (estimated).[30] Total payments to all investors in the same years will be $310 million, $688 million, and $978 million, respectively. Note that these increases occur despite a growth in access charges by local carriers from 16 percent of revenues in 1982 to 50 percent in 1988. Indeed, by 1988, MCI's total payments to all investors are estimated to equal 40 percent of its revenues net of access charges.[31]

It is worth keeping in mind several aspects of these figures. First, MCI's network is just about completed; expansion of the network is expected to be dramatically slower in future years. Second, the capacity of the network is large compared with current use. If sales were significantly better than the projections estimate, costs would rise substantially less than revenues, and cash flow would increase even more. Third, the projections of sales and market share for MCI shown in Table 5-4 are at the low end of the range of industry estimates. In fact, by mid-1988 MCI had already disproved the pessimistic forecasts; its profits had soared, and its stock price had tripled from a year earlier.

The importance of these observations is *not* whether they imply that MCI is a good investment. For our purposes, they shed light on the key issue of whether competition is workable. As long as MCI has a strong cash flow after servicing its debt, it is a viable competitor. Indeed, its extraordinary cash flow makes it a better target for takeover than for bankruptcy.

Much the same story applies to AT&T's other competitors. During the 1980s, they have made massive investments in transmission capacity, much of it in fiber optics, which is now being written off using the generally favorable tax rules regarding depreciation that applied until January 1, 1987. Like MCI, their net profits exaggerate their financial difficulties because they reflect the depreciation deductions for investments in excess capacity. For example, US Sprint has half the business of MCI but two-thirds the capacity, suggesting an even higher ratio of depreciation to revenues. If its depreciation deduction were, say, $300 million in 1986, the company would have had a positive cash flow in the first half of 1986 of about $50 million.

What the new competing carriers have done is to position themselves for a major growth in sales in the late 1980s, as equal access gradually becomes widely available.[32] This does not mean that all will reap lucrative financial rewards in a few years, but it does suggest that because these costs are sunk and because even moderate growth will significantly enhance their cash flows, the other common carriers are likely to be durable competitors.

The best indicator that competition will keep price pressure on AT&T is the relative capacities of the competitors compared with their collective market share. At the end of 1985, AT&T's network had approximately 1 billion circuit miles; the capacity of the competitors was approximately 600 million circuit miles.[33] These figures indicate that the competitors' share of capacity was between 35 and 40 percent—more than double their share of sales. MCI, for example, claims to be capturing between 10 and 15 percent of the business in areas of equal access, but its national share of capacity exceeds 20 percent, indicating its capability to expand sales significantly if AT&T gives it an opening to do so. Moreover, because the competitors do not yet have complete national networks, their relative capacity shares are growing and are generally higher in areas where AT&T faces challenges.

Were it not for the haphazard way that equal access has been provided, combined with the effects of arbitrary LATA boundaries that protect the short-distance toll market of the local exchange carriers, these data would all point to a market that was quite competitive. AT&T's high market share is best explained by a corporate and FCC pricing policy that preserves market share by forcing competitive pricing, combined with the continuing advantage arising from the pace and pattern of equal access.

Our view that the other common carriers have substantial excess capacity implies that their marginal costs of service are far below their current prices. Why, then, do they not cut prices in order to increase market share and capacity utilization? One possibility—to us, unlikely—is that they believe their customers are not very sensitive to price. Another possibility is that they believe AT&T would match any price reduction, leaving their market shares and utilization rates unchanged. They may also believe that if they keep the percentage gap between their own prices and AT&T's constant, and if AT&T is relieved of regulation, AT&T might lead a general increase in prices. These conjectures about the other common carriers' pricing behavior emphasize the oligopolistic nature of today's long-distance market.

Nevertheless, despite the oligopolistic structure of the long-distance market, its performance has been relatively competitive, even before truly equal access has yet arrived (including that by the non-Bell or independent local telephone companies, as well as the recalcitrant BOCs). Consequently, it seems unlikely that the imperfections in the divestiture plan will remain sufficiently important to warrant continued regulatory scrutiny. The new competitive carriers are viable, and any growth in sales in the coming years will go disproportionately to net cash flow, making them stronger still. To the extent that the new competitors continue to show poor profit performance, the cause will be excess capacity—an overly zealous 1980s' expansion program that gives them excessive interest and depreciation expenses compared with revenues. But that situation is not likely to undermine their viability; the capital costs are sunk, and while the cash flow is strongly positive, competition will be an effective force in the market.

The most important conclusion to be drawn from this analysis is that it is time to plan the deregulation of AT&T, with a target of doing so by about 1990. There are two reasons to do so. First, as argued above, AT&T's competitors are likely to do at least as good a job of placing a ceiling on AT&T's prices as is the FCC. This is not to say that AT&T is subject to perfect competition and will earn only normal profits. To the contrary, the best that can be expected in this industry is dominant firm oligopoly with a handful of significant players. Moreover, we expect the advantages of AT&T to persist for years, owing to its first-in advantage, its benefits from the way divestiture has been handled, and its monopoly position in many smaller markets. But these sources of market imperfection are small compared with the position AT&T enjoyed before divestiture, and they are more the type to be tolerated or handled by antitrust than to be subjected to economic regulation.

The second argument for deregulating AT&T in the near future arises from the very fact that it will continue to have a market advantage. Because AT&T did not divest its manufacturing activities, it is in a position to use manufacturing inputs as a means for evading regulation in any case. The only possible consequence of AT&T's vertical integration is that its network and manufacturing activities will be inefficient if the former continues to be regulated. AT&T's incentive to continue to buy its own high-cost equipment evaporates if its long-distance network is deregulated. Indeed, a major potential flaw in the relief sought by the government was that if long distance turned out to be a natural monopoly and continued to be regulated, the failure to divest Western Electric would seriously limit the growth of competition in manufacturing to supply the long-distance system. Fortunately, whether long distance is a natural monopoly is now largely moot. So much investment is in place that the industry will be reasonably competitive for as long a time horizon as one can safely adopt in this rapidly evolving industry. In the next five to ten years, AT&T's market advantages will arise primarily from the way divestiture was implemented. An argument for deregulation is that it will prevent the conversion of these advantages to inefficiencies arising from the remaining vertical relationships within AT&T.

Prospects for competition within the LATAs are not nearly so promising, especially for small customers. There are two basic ways that toll competition could emerge inside the LATAs. One would be to permit facilities-based competition using the local telephone companies' access lines, including a requirement that these companies offer equal access to their competitors. As discussed above, however, state regulators have not permitted this form of competition to come about. Even in the most pro-competitive states, the range of potential competition has been restricted by licensing requirements and the failure to mandate equal access. The second path would be to introduce competition in basic access to the telecommunications system (here defined as all the separately owned but interconnected local and national networks). This path, commonly known as bypass, has proved more promising, at least for large users, because the technologies are controlled largely by the FCC, which has regulated them in a manner that, if not pro-competitive, is at least much less anticompetitive than state regulation. Nevertheless, the scope for using bypass to promote competition in basic access

Table 5-5

Alternative Access for Larger Users

A. Switches	1982 (millions of lines)	1986 (millions of lines)
BOC Business	12	16
BOC Centrex	5.6	5
PBX	15	26

	1982		1986	
B. Lines	Lines (millions)	Revenues (billions)	Lines (millions)	Revenues (billions)
BOC private & special use	NA	$3.6 (1984)	2.8	$4.7
Private microwave	.27	NA	3.4	0.3
Private fiber optic cable	0	0	250	0.2–0.4
Metropolitan area network	0	NA	8	NA
Satellite earth stations	.40	NA	0.8	0.45–0.6

NA = not applicable.

Source: U.S. Department of Justice, Antitrust Division, *The Geodesic Network: 1987 Report on Competition in the Telephone Industry*. Special report prepared by Peter W. Huber. January 1987, chap. 2.

and intraLATA toll is limited by the strong position held by local exchange carriers in terminating MTS calls that originate on bypass facilities.

For large users, the story regarding alternative access to intraLATA toll or other services is provided in Table 5-5. (A relevant backdrop to this and the following table is that local exchange carriers have approximately 115 million switched-access lines in use, and perhaps 40 to 50 million more such lines in unused capacity.) The data on alternative access for large users provide some rather amazing contrasts. First, BOC private and special access lines in use are a drop in the capacity bucket, and BOCs no longer provide half the switching capacity. Second, despite those facts, the BOCs derive more than 80 percent of the revenues from access lines! Even if all the BOCs' unused capacity is assigned to private and special access classes—which is technically incorrect—the installed capacity of private fiber optic cable dwarfs the BOCs' capacity. The principal conclusion to be drawn from Table 5-5 is that although alternative access has finally emerged as important, it has not yet seriously eroded the BOCs' market position. Most likely, alternative access for large users is becoming much like long distance: the local exchange carriers are and will remain the major players, but competitive alternatives will limit BOC prices and promote efficiency of operations, to the extent permitted by state regulation.

Unfortunately, the access alternatives available for large users have thus far solved only half the problem of the bottleneck monopoly. With exceptions such as calls between the plants or offices of large firms that have their own telecommunications networks, these alternatives are used only to bypass local exchange carrier access at the originating end of a telephone call. Typically, interchange service terminates over the access lines of the local exchange carrier. Hence, state regulators can at present preserve the large revenue contribution from bypass customers by setting high termination access charges. Eventually, however, this, too, shall pass, for interexchange carriers could design their networks to terminate their connections over bypass facilities whenever the called party is a bypasser. But doing so will require some reengineering of bypass connections and reprogramming of the interexchange carrier networks—a not inconsiderable cost that the interexchange carriers are unlikely to incur for a long time. Until they do, competitive, cost-based access fees and intraLATA toll charges will not emerge.

For small users, the options are still too limited to produce any serious competition for access and intraLATA toll. Table 5-6 shows the access alternatives for small users. So far, mobile telephones have been the intended use of cellular radio, which makes them a much more expensive alternative form of access because the cellular switching system must be constructed to permit "handoffs" from one cellular area to another as a mobile customer travels. Thus, for plain vanilla telephone service in a

Table 5-6
Alternative Access for Small Users
(Millions of Lines)

	1982	1986
BOC	69	78
Shared tenant service (STS)	0	0.1–0.2
Mobile	<0.3	<0.5

Source: Huber, *Geodesic Network*, 2.12.

residence or small business, cellular is not currently a realistic alternative unless the customer places a high value on mobility. In the future, however, fixed radio telephone technology has obvious potential to provide economical local access service to at least some fixed local telephone customers, especially residential users with long loops. But whether the technology will be permitted to develop in that direction while largely under the control of wireline carriers is doubtful.

Another potential alternative for local access is "shared tenant service" (STS), which is simply another name for a local private network. As with many other services, the distinction between private networks and shared tenant services is essentially an institutional one to aid in discrimination in regulation and pricing. The idea is for a group of separate customers to band together to buy a switch (usually a PBX) and connect all their telephones and other communications equipment to it. They can then negotiate with various suppliers of access lines: the BOCs for local service, and a larger number of alternatives for long-distance interconnection. The advantage is that if the number of users is large enough, they collectively exercise the options available to the "large user" (see Table 5-5). Among other things, this arrangement enables them to use private lines as a basis for bypassing the local carriers for all toll service within or between LATAs. (Another benefit is that they need fewer total access lines because, statistically, they are unlikely all to be calling simultaneously.)

Shared tenant service is obviously a significant threat to the local exchange carrier monopoly. As switches (like other computers) fall in price, the minimum number of users who can profitably use the technology continues to drop. And if customers were to pick up this option in significant numbers, the demand for both local exchange access lines and intraLATA toll services could be severely affected. Accordingly, it is not surprising that state regulation of shared tenant service is generally highly restrictive. Table 5-7 shows the general pattern of restrictions, which might "seem superficially innocuous but in fact impede the functionality and raise the price of STS services."[34]

Until state regulatory policy either permits access alternatives for small users or requires equal access for intraLATA competitors over local networks, there is really no hope that serious competition will emerge inside the LATAs. Our expecta-

Table 5-7
State Regulation of Shared Tenant Services (STS)

AMERITECH	IL	IND	MI	OH	WIS		
	D	D	D,S	D	S		

BELL ATLANTIC	DEL	D.C.	MD	NJ	PA	VA	WV
	D	N	D,N	D	D	D	D,N

BELLSOUTH	AL	FL	GA	KY	LA	MISS	NC	SC	TENN
	D,C	P,C,S	D,S	C	(D,S)	D	P,C	D,S	(D,S)

NYNEX	CONN	ME	MA	NH	NY	RI	VT
	D,N	#	(D),C,(S)	#	D,C	#	D,C

PACIFIC	CA	NV
	D,S,N	D,N

SOUTHWESTERN	AK	KAN	MO	OK	TX
	P,C	(P,C),S	A,S	(P,C,S)	(P),D,C(S),N

US WEST	AZ	CO	ID	IO	MINN	MONT	NEB
	D,S	D,N	D	D	D,N	D	D,N
	NM	ND	OR	SD	UT	WASH	WYO
	D,C,S	(D)	(C)	Yes	D	(D)	C,S,N

P: STS systems are prohibited unless *partitioned*.
D: STS systems are charged *discriminatory* rates by comparison with single-user PBXs (such discrimination can take the form of (i) higher rates for local services, (ii) additional charges for each user of the system, or (iii) denial to STS systems of flat rate options available to single-user PBXs).
C: STS systems must be *certificated* by the state regulatory commission in all or some instances.
A: Tenants in an STS building are not entitled to receive direct *access* to telephone company service.
S: STS systems are limited in *scope* by narrow ownership, geographic, or trunking restrictions.
N: STS systems must pass through local service costs on a *nonprofit* basis.
Yes: State regulatory commission authorizes resale of local service by STS providers without any of the above restrictions.
#: No action has been taken regarding STS and the status of STS arrangements under older tariffs is unresolved.
( ) Restrictions are only proposed at present, and are awaiting action by the state regulatory commission.

Source: Huber, *Geodesic Network*, Table L.10.

tion is that—as occurred at the federal level—over an excruciatingly long period, state regulators will gradually relax their prohibitions against competition. Although there is no reason to be optimistic about the speed of this transition, the federal government might change the picture if it adopts a more aggressive pro-competitive stance than has emerged in either the FCC's regulations or the Department of Justice's reports about the rules of divestiture.[35] The tools would be (a) continued use of the technologies under FCC jurisdiction to undermine anticompetitive policies in the states and (b) use of the "waiver process" whereby BOCs are permitted into competitive telecommunications markets as a lever to promote competition with franchised BOC monopolies. An example of the first would be to allocate more spectrum to radio telephony, license more radio telephone competitors in each metropolitan area, and permit the use of cellular frequencies for fixed (rather than mobile) service. Another example would be to assert jurisdiction over the licensing of switches that, in part, are connected to lines in the FCC's jurisdiction (such as bypass facilities). An example of the second would be to permit BOC entry into, say, PBX manufacturing only if shared tenant services were totally deregulated in the service territory or to permit BOC entry into long-distance interconnection services only if there were open entry and equal access for intraLATA toll.

These policies may well prove ineffective in breaking open the LATAs to effective competition. State regulators and local exchange carriers may conclude that a limited local monopoly beats a larger competitive environment and so may respond to these policies by protecting as much of the intraLATA monopoly as possible. If so, the path to intraLATA competition will be long, and the disparities in costs and quality of service between large and small users will continue to grow.

Telephone Equipment

AT&T long monopolized the manufacture and sale of telephone equipment in the United States, especially in the vast geographic territories served by its local operating companies. In contrast to the markets for long-distance service, those for telephone equipment have rapidly become deconcentrated since the introduction of competition.

Although several distinct categories of telephone equipment are marketed, the most important distinction is between equipment used by customers (called terminal equipment or customer premise equipment [CPE]) and equipment used by telephone companies in their central offices or transmission facilities. CPE is often further subdivided, at least for statistical purposes, into (a) single-line equipment, mainly purchased by residential customers, and (b) multiline equipment, such as PBXs and key sets, mainly used by businesses.

Under the *Carterfone* decision, the FCC in 1968 permitted entry into the provision of CPE; however, AT&T continued to struggle to maintain its monopoly by insisting on the use of protective devices to shield the network from potential harm. In the mid-1970s the FCC removed this last institutional barrier to entry by outlawing the protective device requirement and substituting its own type-acceptance program for telephone equipment. As a result, and perhaps because Western Electric was (relative to Long Lines) especially inefficient as a supplier, AT&T's share of U.S. sales of CPE fell rapidly.[36] In 1968, at the time of *Carterfone,* that share must have been approximately 100 percent of equipment sales of each type[37] within Bell franchised territories, translating to as much as 85 percent of all U.S. telephone equipment sales.[38] By 1983, just before divestiture, AT&T sales of business (multiline) CPE had declined to about 58 percent of U.S. sales (including all revenues from Centrex service). By 1986, with all Centrex revenues accruing to the divested BOCs, AT&T's share of multiline CPE had fallen to about 16 percent, while the BOCs (via Centrex) had 42 percent. As Table 5-8 shows, AT&T's share of U.S. PBX sales, excluding Centrex, was around 20 percent in 1986.

A similar story of rapidly declining market share characterizes the remaining equipment categories. In 1986, for example, AT&T's U.S. share of handsets was 36 percent; of key systems, 25 percent; of digital central office switches, 49 percent; of metal cable, 50 percent; of microwave equipment, 17 percent; and of fiber optic cable, 36 percent. (See Tables 5-9–5-14.)

One remarkable change in residential CPE was the switch from leasing to owning telephones. In 1982, 90 percent of consumer telephones were leased from AT&T. As a result of AT&T's marketing campaign and the prices at which it offered to sell equipment in place, 70 percent of all residential telephones were consumer owned in 1986. U.S. retail sales of CPE to consumers are now about $1.75 billion annually.

AT&T undertook drastic measures to reduce costs, mainly in the equipment area, in response to its declining market share. Its overall postdivestiture work force was about 380,000 in January 1984; by early 1987, AT&T's work force had been reduced by about 80,000 jobs. (See Table 5-15.)

Table 5-8

PBX Manufacturer U.S. Shares, 1982–1986

(Percentage of Total Lines Shipped)

	1982	1983	1984	1985	1986
AT&T	25–29	21–25	17–21	20–26	20–23
Northern Telecom	13–14	16–18	21–23	22–23	23
Rolm (IBM)	12–15	13–17	16–18	14–15	15–18
Mitel	12	10–13	9–11	8–10	9–10
NEC	4–5	4–6	6–7	7	7–9
GTE	5	4–5	3–5	3–5	4
Siemens	4	4–5	4	2–5	4
Intecom	1	2–3	3–4	3–4	2
Fujitsu	3	3	2	2	2
Harris	2	2	2–3	2–3	2
Ericsson	1	2	2–3	2	2
Other	9–18	0–19	0–15	0–15	1–10
Total lines shipped (millions)	3.2	3.7	4.4	4.8	4.9–5

Note: Excludes Centrex revenues.

Sources: Huber, *Geodesic Network,* Table PX.4, 16.4, citing Gartner Group, *Local Area Communications: PBX Market Year-End Review* (February 26, 1986); Eastern Management Group, PBX: *New Environment* 106 (January 1986); Probe Research, *Probe on PBX* 319 (1984); The Yankee Group estimates; "Competition Is Tough, Margins Are Slim, but Vendors Still Flock to PBX Market," *Communications Week* C1 (April 1, 1985); Goldman, Sachs Research; and Ameritech submission (June 12, 1986) (citing Eastern Management Group).

Table 5-9

Handset Manufacturer U.S. Shares, 1986

(Percentage of Total Lines Shipped)

Manufacturer	Share (%)
AT&T	36
ITT	9
Conair	7
GTE	7
Radio Shack	6
Panasonic	6
Cobra	3
Uniden	2
GE	2
Sanyo	2
Sony	2
Telemax	2
Webcor	2
Others	16

Source: Huber, *Geodesic Network,* Table 5, 17.4, citing Ameritech (September 12, 1986) submission, which cites The Yankee Group, 1986.

Table 5-10

Key System Manufacturer U.S. Shares, 1985

(Percentage of Total Lines Shipped)

Manufacturer	Share (%)
AT&T	25–26
TIE	16–24
ITT	8–17
Iwatsu	4–7
Vodavi	5
NEC	4–5
Toshiba	3–9
Others	7–35

Source: Huber, *Geodesic Network,* Table T.7, 17.6, citing The Yankee Group, as presented in *Communications Week* C2 (December 23, 1985); Northern Business Information, as presented in *Communications Week* 33 (July 7, 1986); and Dataquest Incorporated, *Telecommunications Industry Conference* 12 (1986).

Table 5-11

Manufacturer U.S. Shares of Central Office Switches

(Percentage of Lines Shipped Annually[a])

	U.S.		World	
	1982[b]	1986[c]	1982[d]	1986[e]
AT&T	<1/70	49	<1/	21–26
Northern Telecom	66/17	35	29/	25
GTE	3/7	10	2/	8–9
CIT-Alcatel		<1	41/	13–14
Ericsson			14/	8
NEC		<1	5/	7–8
ITT	/2	<1		3–4
Siemens				5–6
Plessey/Stromberg	/2	2	5/	1
Others	31/2	3	4/	0–9

a. For 1982, the first market share shown is for digital switches alone, and the second market share is for digital and analog switches combined; 1986 numbers are for digital lines only.

b. Source for digital market shares: AT&T submission (citing Northern Business Information). Source for analog and digital market share combined: "CO Switch Vendors Eye Battle for BOC Market—Newcomers Say There Is Room for More than Two Suppliers to BOCs," *Communications Week* C1 (August 27, 1984) (1983 market shares).

c. Source: Northern Business Information estimates.

d. Source for digital market shares: AT&T submission (citing Northern Business Information).

e. Sources: Memorandum from Lexecon Inc. to John Thorne, Esq. (April 28, 1986), *Assessment of GTE Activities and Performance* (citing Gartner Group and Northern Business Information); and AT&T submission (citing Northern Business Information) (1985 data).

Source: Huber, *Geodesic Network,* Table CO.4, 14.7.

Table 5-12

Manufacturers' U.S. Shares of Transmission Equipment

	Metal Cable[a]	Microwave[b]	Fiber[c]	Satellite[d]
1. Firm Shares of Revenues				
AT&T	50	17	36	
General Cable	12			
Ericsson	8			
Essex	7			
Superior Cable	7			
Celwave	6			
Rockwell International		43		
NEC		14		
Northern Telecom			17	
Corning			32	
ITT			5	
Harris/Farnion		6		32–40
Scientific Atlanta				15–20
M/A Com				15–27
Satellite Transmission Systems				11
General Instrument				9
California Microwave				8
Equatorial				5
Others	11	20	10	0–8
2. Total Revenues (Billions of Dollars/Year)	1.1	0.3–0.5[e]	1.1	1.6[f]

a. 1985 data. Sales to telcos only.
b. 1984 data.
c. 1986 data. Without associated electronics.
d. 1984 market shares; 1986 figures may be significantly different.
e. $0.5 billion is an estimate for 1985.
f. Earth stations only. Space satellites account for another $2.2 billion in sales.

Source: Huber, *Geodesic Network,* Table M.3, citing Kessler Marketing Intelligence estimates; U.S. Department of Commerce, Office of Telecommunications estimates; Ameritech (October 6, 1986) submission, which cites International Data Corporation, *Microwave Market* 8, pp. 14–17 (1985); U.S. Department of Commerce, *U.S. Industrial Outlook 1986* 29-6 (1986); "Local Bypass Stimulates Microwave Sales," *High Technology* 26 (May 1986); The Yankee Group, *Carrier Systems* 11 (December 1984); and Northern Business Information, *Transmission Equipment Market 1986 Edition* 53 (June 1986).

Table 5-13

Mobile Terminal Manufacturer U.S. Shares, 1986

Manufacturer	Cellular Sets[a] (%)	Paging Units[b] (%)
Motorola	14	80–85
OKI	14	
Audiotel	11	
NEC	11	5–8
Panasonic	10	4–6
Novatel	10	
Mitsubishi	7	
General Electric	6	
Others	17	5–10

a. Percentage of 1986 first-quarter sales.
b. Estimated percentage of cumulative units shipped through the end of 1986.
Sources: Huber, *Geodesic Network,* Table T.6, 7.5, citing Ameritech submission, which cites Herschel Schosteck Associates (for cellular sets) and the Eastern Management Group for paging units.

Table 5-14

Manufacturer Shares of Transmission Equipment
Sold to U.S. Telephone Companies
(Percentage of Revenues from Sales to U.S. Telcos, 1985)

Manufacturer	Share (%)
AT&T	49
Rockwell	7
Northern Telecom	6
NEC	4
General Cable	3
Ericsson	3
Siecor	2
ITT	2
Lynch	2
R-Tec	2
GTE	2
Others	18

Source: Huber, *Geodesic Network,* Table M-2, citing Northern Business Information, *Transmission Equipment Market* 8 (June 1986).

Table 5-15
AT&T Employment Levels

	AT&T	*BOCs*	*Total*
December 31, 1980	1,044,000	—	1,044,000
January 1, 1984	384,000	587,000	971,000
December 19, 1986	290,000[a]	547,000	837,000

a. Announced target for 1987.
Sources: *Moody's Public Utility Manual 1981,* vol. 1, p. 71, and Barnaby J. Feder, "AT&T Will Cut 8.5% of Staff: 27,400 Affected; Write-down of $3.2 Billion Set," *New York Times,* 19 December 1986, sec. D.

Developments in the central office equipment business have differed from the CPE story. Central office switches and other equipment used by telephone companies have traditionally been purchased from captive, vertically integrated suppliers.[39] The *Carterfone* decision did not affect the incentive or ability of BOCs to buy Western Electric equipment for their own use, and a similar pattern prevails in other industrialized countries. (See, e.g., Table 5-16.) The postdivestiture environment is the first in modern times in which large telephone companies have not been integrated. Northern Telecom, however, preceded AT&T as a large equipment manufacturer whose captive customer base was far smaller than its capacity.

Table 5-16
Telco Switch Purchases from Affiliated Manufacturers
(Percentage)

Bell System	
1982	>95
1985 (from AT&T)	50
GTE (1985)	87
Bell Canada (Northern Telecom) (1984)	100[a]
DBP (Siemens) (1984)	60[b]
France (Alcatel) (1984)	80
Japan (NEC) (1984)	(NA)[c]
Sweden (Ericsson) (1984)	100
Britain (Plessey/GEC) (1985)	<60

a. In areas served by Bell Canada.
b. ITT is the second preferred vendor in West Germany and accounts for 40 percent of the market.
c. NEC has a historical preferred vendor status in Japan, but NTT recently signed a substantial contract to purchase switches from Northern Telecom.
Sources: Huber, *Geodesic Network,* Table CO.6, 14.9, citing Arthur D. Little, cited in *Financial Times,* January 6, 1986, p. v; Northern Business Information, *Central Office Equipment Market* 81 (1986 Edition); and *Invasions and Counterinvasions, IEEE Spectrum* 66 (November 1985).

Table 5-17
Trade in Telephone Equipment (SIC 3661)
(Millions of Current Dollars)

Year	Value of U.S. Production	Imports	Exports
1972	3,974	86	77
1975	4,734	93	198
1980	11,162	421	577
1986[a]	18,213	2,185	850

a. Estimates.
Source: *U.S. Industrial Outlook* 1987 (Washington, D.C.: Government Printing Office, 1987), pp. 30-1 and 30-8.

The divestiture has lessened the incentive of the BOCs to purchase from Western (now AT&T Information Systems), a fact consistent with the government's contention at trial regarding the effects of AT&T's vertical integration on this incentive. (See Table 5-16.) Imports from Europe, Japan, and Canada have begun to challenge AT&T's monopoly of this field, and for this and other reasons the United States now has a massive trade deficit in telephone equipment. (See Table 5-17.) Nevertheless, the domestic market, especially the market for central office switches, is highly concentrated. Only Northern Telecom has succeeded in taking significant BOC business away from AT&T (see Table 5-18). And vertically integrated GTE still buys little from others.

The failure of European and Japanese firms to make more serious inroads into BOC sales is matched by AT&T's failure to sell equipment abroad. AT&T's recent attempts through joint ventures with foreign manufacturers to penetrate world markets for central office equipment have not yet borne fruit. AT&T's worldwide share of central office equipment sales is estimated at about 12 percent, well behind Northern Telecom's 22 percent. (See Table 5-19.) In part, AT&T's problems may stem from policies in other countries (where telephone networks are typically government owned) to prefer domestic suppliers. But the similarly slow penetration of the U.S. market by foreign firms (except Northern Telecom) suggests that technical differences in network design and operations between North America and other advanced countries may also constitute a barrier to entry both in and out of the United States.

Table 5-18
Buyer Preferences of Switches Among Manufacturers, 1985
(Percentage of Purchases from Various Manufacturers)

	Manufacturer			
	AT&T	*NTI*	*GTE*	*Other*
Ameritech	55	45	0	0
Bell Atlantic	79	21	0	0
BellSouth	57	43	0	0
NYNEX	4	96	0	0
Pacific Telesis	6	94	0	0
Southwestern Bell	77	23	0	0
US West	51	49	0	0
GTE	3	10	87	<1
Other, U.S.	20	46	5	30

Source: Huber, *Geodesic Network*, Table CO.8, 14.10, citing Northern Business Information, *Central Office Equipment Market* 81 (1986 Edition).

What conclusion can be drawn from these data on trends in the telephone equipment business? Although it is probably too soon to assess the impact of divestiture on equipment markets, the following conjectures seem to be supported by the available data:

- The CPE market was on its way to being structurally competitive as a result of the FCC's actions. Divestiture, while it may have sped things along, was arguably unnecessary to the emergence of healthy competition in this market.
- Western Electric was an extremely inefficient, high-cost supplier of telephone equipment; otherwise, its market share would not have fallen so far so quickly as a result of new entry and of divestiture.

Table 5-19
Worldwide Share of Sales of Central Office Equipment, 1985

Northern Telecom	22%
AT&T	12%
Alcatel	12%
GTE	10%
Ericsson	8%
Siemens	8%

Source: Northern Business Information, in Karen Lynch and Fredric Paul, "Central Office Vendors Vie for the Promised Land," *Communications Week,* October 13, 1986.

Table 5-20
Purchases of Transmission Equipment

	Metal Cable[a]	Microwave[b]	Fiber Cable[c]	Satellite[d]
1. Buyer Shares of Total Outlays				
BOCs				
Single	8–16	4[e]	3–13	0
Total	81	25	46–57	0
Independents	19	11	6–14	20
ICs		41	10–28	
Nontelco Buyers		23	10–20	80
2. Total Annual Outlays				
($ billions)	1.1	0.3–0.5[f]	1.1	1.6[g]

a. 1985 data. Sales to telcos only.
b. Based on forecasted 1986 outlays.
c. 1986 estimates. Ranges based on estimates by outlays and by kms. of fiber purchased.
d. 1984 data. Includes only transmit and receive earth stations; nontelco buyers purchase about 95 percent of the smaller receive-only earth station market.
e. Single BOC share derived by dividing BOC total by seven.
f. $0.5 billion represents an estimate for 1985.
g. Earth stations only. Space satellites account for another $2.2 billion in sales.
Source: Huber, *Geodesic Network,* Table M.1, citing Northern Business Information, *Transmission Equipment Market* 58 (June 1986) and Ameritech (October 6, 1986) submission which cites International Data Corporation, *Microwave Market* 8, pp. 14–17 (1985).

- Centrex, to the extent that it was a problem before divestiture, remains one today.[40] One good reason to let BOCs sell PBX equipment is that their incentive further to distort Centrex pricing is reduced. In effect, because of Centrex, BOCs are in the PBX business willy-nilly but through an expensive technology that distorts pricing and network design.
- The markets for equipment used by telephone companies have become more competitive since divestiture, and divestiture was probably necessary to achieve this change. Buyer concentration in these markets has been reduced substantially. (See Tables 5-20–5-26.) Seller concentration is still high; however, the lower prices and costs after divestiture indicate that a tight oligopoly can outperform a regulated, vertically integrated monopoly.

One of the policy issues facing the government today is whether, as the Huber report[41] and the Department of Justice

Table 5-21
Buyer versus Seller Shares in Central Office Equipment
(Percentage of Lines Purchased and Lines Supplied, U.S. Market)

Telcos and Other Carrier Switch Purchasers	*Access Lines Purchased (%)*	*Access Lines Supplied (%)*	*Manufacturer*
1. 1982[a]			
Bell System	17/80	<1/80	Bell System
GTE	6/10	3/10	GTE
Others	77/10	96/10	Others
2. 1986[b]			
Ameritech	12	49	AT&T
Bell Atlantic	17	35	Northern
BellSouth	15	10	GTE
NYNEX	9	6	Others
Pacific Telesis	9		
Southwestern	4		
US West	6		
GTE	14		
Other, U.S.	14		

a. The first figure listed represents share for digital switches only, while the second figure represents share for both digital and analog switches.
b. Buyer shares are for 1985.
Source: Huber, *Geodesic Network,* Table CO.5, citing Northern Business Information, *Central Office Equipment Market* 81 (1986 edition), AT&T submission.

Table 5-22
Buyer versus Seller Shares: Cellular Systems, 1986
(Percentage of Systems Contracted or On Line, Top Ninety U.S. Markets)

Cellular System Operator	*Systems Purchased[a] (%)*	*Systems Supplied (%)*	*Manufacturer*
Cellular One	19	40	Motorola
GTE Mobilnet	12	32–36	AT&T
Ameritech Mobile	7	13–15	Northern/GE
NYNEX Mobile	7	4–10	Ericsson
Southwestern Bell Mobile	7	8	Others <5% Each
Bell Atlantic Mobile	6		
BellSouth Mobility	6		
NewVector Communications	4		
PacTel Mobile Access	3		
United TeleSpectrum	3		
Other < 3% each	26		

a. Calculated from percentage of installed cells sites of cellular providers in top ninety markets. Does not reflect mergers and acquisitions since June 1986.
Source: Huber, *Geodesic Network,* Table CO.5, citing "Cellular Business," *The Journal of Cellular Telecommunications* (June 1986).

Table 5-23
Buyer Shares of Central Office Switches
(Percentage of Digital Lines Purchased)

	United States		World	
	1982[a]	1985	1982	1985
Ameritech	]	12	]	5
Bell Atlantic	]	17	]	8
BellSouth	]	15	]	7
NYNEX	]17/80	9	]40	4
Pacific Telesis	]	9	]	4
Southwestern Bell	]	4	]	2
US West	]	6	]	2
GTE	6/10	14	/	6
Other, U.S.	77[b]/10	14	/	6
Other, World			60	56

a. The first figure listed under 1982 represents share for digital switches only, while the second figure represents share for both digital and analog switches.
b. Includes purchases by REA.

Sources: Huber, *Geodesic Network,* Table CO.3, citing Northern Business Information, *Central Office Equipment Market* 81 (1986 edition); AT&T submission, "The Local Exchanges Remain Bottlenecks for All Manufacturers of Telecommunications Equipment," fig. 90, which cites Northern Business Information; "The World War in Central Office—Survival Needs Drive Switch Makers to Seek Global Expansion," *Communications Week,* C1 (November 18, 1985).

Table 5-24
Central Office Digital Switching: Largest Purchasers
by Disbursements, 1984

Purchaser	$Millions
DGT (France)	371
GTE	312
BellSouth	105
NTT (Japan)	85
British Telecom	84
NYNEX	83
Pacific Telesis	80
US West	36
Ameritech	33
Southwestern Bell	33
Bell Atlantic	27
Bundespost (Germany)	24

Sources: Huber, *Geodesic Network,* Table CO.3, citing Arthur D. Little; Northern Business Information.

Table 5-25
Buyer versus Seller Shares: Microwave
(Percentage of Dollar Outlays or Revenues, 1986–1984)
(U.S. Market)

Buyers[a]			Sellers[b]
BOCs	25	43	Rockwell Int'l.
Independents	11	17	AT&T
AT&T	16	14	NEC
Other ICs	25	6	Harris/Farinon
Nontelco	23	20	Other

a. Based on forecasted 1986 outlays.
b. Based on 1984 revenues, and so exclude most of AT&T's and other manufacturers' short-haul microwave offerings.

Source: Huber, *Geodesic Network,* Table M.8, 15.9, citing Ameritech (October 6, 1986) submission, which cites International Data Corporation, *Microwave Market* 8, pp. 14–17 (1985).

Table 5-26
Buyer versus Seller Shares: Fiber Optic Cable
(Percentage of Dollar Outlays or Revenues, 1986, U.S. Market)

Buyers			Sellers
Ameritech	5–8	36	AT&T
Bell Atlantic	5–8	32	Corning
BellSouth	10–11	17	Northern Telecom
NYNEX	4–6	5	ITT
Pacific Telesis	3–8	10	Others
Southwestern Bell	10–13		
US West	5–7		
Independents	6–14		
AT&T	2		
Other, U.S.	28–46		

Note: Ranges for buyers based on estimates by outlays and by kms. of fiber purchased. Expenditure shares computed using Kessler Marketing Intelligence estimates for the total market. Expenditure share for ICs and nontelcos equals the residual. Shares of fiber km. sold from BOC submission that cites Kessler Marketing Intelligence and various BOC annual reports. Sellers' shares based on revenues from sale of fiber.

Sources: Huber, *Geodesic Network,* Table M.7, 15.8, citing "Expenditures for BOCs and Independents" from "Carriers Plan $24 Billion Outlay in 1986," *Telephony* 32–40 (January 13, 1986), and BOC submission that cites Kessler Marketing Intelligence estimates.

recommend, to remove the restrictions on BOC entry into the manufacturing of telephones and other CPE. Judge Greene has already reduced some of the theoretical effectiveness of divestiture by permitting the BOCs to sell CPE purchased from others. Would entry into manufacturing make any difference? The evidence from the predivestiture period is that even geographically "small" telephone companies, such as GTE and European systems that manufacture equipment, succeed in monopolizing the equipment business in their territories. GTE certainly did in the period before *Carterfone*.[42] On the other hand, the FCC's equipment type-acceptance and registration program seems to have been effective in opening entry to independent CPE vendors, for customers of AT&T and other telephone companies. It therefore seems likely that the difficulties of cross-subsidization and discrimination are sufficiently great in the CPE business that joint ownership of local exchange facilities and equipment businesses will not lead to complete monopolization. Whether it nevertheless will lead, as theory predicts, to an inefficiently large CPE market share for local telephone companies is possible but cannot be proven without experimenting with a relaxation in the restrictions. The Department of Justice, in its recent triennial report to the court, takes the economically inexplicable but politically expedient view that, under the MFJ and the antitrust laws, such a reduction in economic efficiency is a matter for concern only if it leads to actual monopolization.[43]

The analysis of markets for central office equipment and other capital goods sold to telephone companies is more clear-cut. Permitting BOC manufacture of such equipment will probably destroy the nascent competition in this industry. The evidence from the predivestiture period, including the evidence from the behavior of large non-Bell operating companies, is that telephone companies gain from captive equipment sales. Although it has not been proved, the presumption is that this behavior is inefficient and driven in part by the perverse incentives of rate-base, rate-of-return regulation as discussed above.

Extensive self-dealing in equipment is also consistent with strong economies of vertical integration between the manufacture of central office and transmission equipment and ownership of local exchange facilities. One way to test this hypothesis is to ask whether, in the period since divestiture, the BOCs have entered into long-term contracts or other near-substitutes for

Table 5-27
BOC Operating Income from Regulated and Unregulated Operations, 1985
(Billions of Dollars)

BOC	Regulated Telephone and Yellow Pages	Unregulated Subsidiaries
Ameritech	2.42	−0.135
Bell Atlantic	2.45	−0.126
BellSouth	3.20	−0.072
NYNEX	2.59	−0.280
Southwestern	2.20	−0.075
Pacific	2.26	−0.006
US West	2.24	−0.245

Source: Huber, *Geodesic Network,* Table PX.15, 16.28, citing "Comments of the North American Telecommunications Association Regarding the Continuing Need for the Manufacturing Prohibition of the Modified Final Judgment," app. D (June 2, 1986).

vertical integration with equipment suppliers. If there are economies of vertical integration in this area, they may be achieved through long-term contracts, perhaps supplemented by joint R&D projects, which have less chance of permitting anticompetitive behavior than vertical integration does.

A final piece of relevant evidence regarding BOC participation in competitive markets is their actual experience in the markets that Judge Greene permitted them to enter. The theory of the AT&T case predicts that the BOCs will enter markets that are not profitable as a way of preserving or expanding sales and rate base and of thwarting the growth of actual or potential competitors for various local services. The response of AT&T to competition, which included massive cost-reduction efforts, suggests that the BOCs' performance in competitive markets will be further inhibited by a high cost structure inherited from predivestiture AT&T. In fact, both predictions are borne out by the financial performance of BOCs in competitive markets, which apparently has been dismal thus far.[44] (See Table 5-27.)

Conclusions

The early experience after divestiture leads to several conclusions about the validity of the original concept of *United States v. AT&T.* The primary purpose of the case was to introduce competition in equipment manufacturing and long-distance service without structurally creating competition in either case. Not surprisingly, in a few years these markets have been transformed only from monopolistic ones to oligopolistic ones, with the latter ranging from a fairly competitive structure for CPE to a dominant firm in long-distance, and close to a duopoly in central office switches. Nevertheless, in each case the market position of AT&T has eroded significantly, and prices and costs have fallen. Although competition may be far from perfect in some of these markets, it nonetheless has proved sufficient to outperform the combination of regulated monopoly and vertical integration that preceded it.

The early experience also suggests a continuing set of policy problems that are unlikely to disappear quickly. Specifically, local exchange carriers are still regulated monopolies in a wide range of services and are desirous of expanding their lines of business into additional areas that are competitive and unregulated. The tension that continues to plague policymakers is the traditional one: the possibility of efficiency advantages from vertical integration and multiproduct production versus the efficiency losses arising from regulated monopoly (including inefficient self-dealing with related enterprises and anticompetitive strategies such as internal subsidization and raising rivals' costs). Behind this dilemma stands the incompatibility between federal and state regulatory policies, with the latter generally very reluctant to permit competition into protected monopoly markets of the local exchange carriers.

To some degree, the magnitude of this policy problem has been made greater by the details of divestiture and by the policies of the FCC. The anticompetitive policies of the states would be a less serious problem if LATAs were smaller, if cellular technology were more directly competitive to local exchanges, and if the FCC were more actively promoting bypass with the technologies under its control. Nevertheless, the basic policy problem is largely political and legal, having to do with the very institution of federalism as it applies to regulated utilities. As long as the last leg of the telecommunications network that connects customers to the system is controlled by state policy, the ultimate hope for maximizing the reach of competition (and, eventually, deregulation) lies in changing the policies of state regulators. In our view, further steps in implementing divestiture (including policies regarding BOC diversification) should more self-consciously take into account the importance of influencing the future direction of state regulation.

Notes

1. A fifth new technology, fiber optic transmission, is of great significance today but had no large effect on the industry before divestiture.

2. "Pair gain" refers to new electronic technologies for using a single pair of copper wires to provide numerous separate local connections to customers.

3. Analysis of the subsidy involves not only the allocation of various non-traffic-sensitive common costs at the local level but also the evaluation of backflows: license fees as well as payments by the operating companies to AT&T for equity capital and for equipment. Peter Temin and Geoffrey Peters have argued that the subsidy did flow as AT&T claimed. See "Is History Stranger than Theory? The Origin of Telephone Separations," *American Economic Review* 75 (1985), pp. 324–327. The subsidy issue, however, is far more complex than simply whether long distance subsidized local service. First, the meaningfulness of the "subsidy" is undermined if (a) local telephone companies have paid supercompetitive prices to vertically related affiliates for equipment or services and (b) regulated monopolies are generally inefficient. Second, further disaggregation of customer classes reveals that the distribution of the alleged subsidy is not uniform across all forms of local service. The best available data, which are crude, suggest that prices for local service are not far from cost for urban residential customers; are far below cost for rural service and, most likely, for business Centrex customers (see Note 40 below); and are above cost for single-line business service as well as long-distance service. See Nina W. Cornell and Roger G. Noll, "Local Telephone Prices and the Subsidy Question," working paper, Stanford University.

4. See Charles Jackson and Jeffrey Rohlfs, "Improving the Economic Efficiency of Access Charges" (Washington, D.C.: Shoshan and Jackson, Inc. 1986).

5. The historical story, of course, was the other way around. AT&T acquired its second monopoly of local exchange companies by denying competing and other independent local exchange companies access to its patent-based monopoly of long-distance service, thus destroying them or forcing them to sell out. See *Report of the Federal Communications Commission on the Investigation of the Telephone Industry in the United States,* made pursuant to Public Resolution no. 8, 74th Congress, pp. 130–131.

6. For discussions of the tendency of a rate-of-return–regulated monopolist to expand inefficiently into related markets, see H. Averch and L. Johnson, "Behavior of the Firm Under Regulatory Constraint," *American Economic Review* 52 (1962), p. 1052; R. Posner, "Natural Monopoly and Its Regulation," *Stanford Law Review* 21 (1969), p. 615; O. Grawe, *Intrafirm Subsidization and Regulation,* FTC Bureau of Economics working paper 91 (1983); O. Williamson, *Markets and Hierarchies* (1975), pp. 113–115; F. Warren-Boulton, *Vertical Control of Markets* (1978), pp. 40–48; and T. Brennan, *Regulated Firms in Unregulated Markets: Understanding the Divestiture in* United States v. AT&T, U.S. Dept. of Justice, EAG discussion paper 86-5 (1986).

7. Later, the D.C. Circuit Court of Appeals, which generally reviews FCC actions, was also an important cheerleader for increased competition.

8. The principal exception to this generalization is R&D, which is especially difficult to attribute to geographic areas and/or categories of services.

9. It would not be quixotic to contend that a policy threatening AT&T with entry if it departed from cost-based prices would be highly desirable, even if AT&T is assumed to be a natural monopolist. Nevertheless, to base an antitrust case on this contention is tantamount to arguing that the Department of Justice has a better theory of regulating prices than the FCC does. While this may be accurate as a matter of economic analysis, it is a legally dubious basis for a case.

10. The existing allocations of electromagnetic spectrum, together with the positioning of cellular radio service as a mobile telephone service, result in radio telephony that is a serious threat to wire distribution only for local wireline exchange customers with relatively long loops, and only then if telephone prices to such customers reflect costs of service. To make radio a broader competitive threat would require, under existing technology, additional allocations of (say) unused UHF television channels and a marketplace repositioning of radio telephony toward fixed-location service. Wireline telephone companies are less likely than independent firms to orient radio telephone service as a substitute for wireline service.

11. See the Huber report at 4.9 and Figure MB.3. The FCC had previously created in each city two allocations of the frequency spectrum for use in providing cellular radio telephone service. One of these allocations in each city had been reserved initially for a "wireline" telephone company, which in most cities, through agreements with the independent telephone companies, was AT&T. The second "nonwireline" allocation was initially reserved for nontelephone-company applicants, but the initial restriction has now expired.

12. See the letter from the Hon. William F. Baxter to the Hon. Patrick J. Power, October 20, 1983, entered in the record of the California Public Utility Commission proceeding OII 83-0601.

13. If the LATAs were much smaller, of course, competition still might not emerge in intrastate, interLATA markets, because state regulators might prevent it or set such high intrastate access charges that competition could provide no significant benefits. Smaller LATAs should be viewed as potentially, but not necessarily, beneficial to competition.

14. Indeed, according to Assistant Attorney General Baxter, "the Department fully expected that state regulatory commissions might allow intraLATA competition and require equal access arrangements for all intraLATA carriers." Moreover, in large LATAs with multiple population centers, "a regulatory environment based on the presumption of monopoly service between such population centers 'would be alien' to the analytic approach used by the Department in making its LATA recommendations." Letter to Power, Note 2 above, pp. 8–9.

15. U.S. Dept. of Justice, Antitrust Division. *The Geodesic Network: 1987 Report on Competition in the Telephone Industry*. Special report prepared by Peter W. Huber, January 1987, 3.16.

16. Huber, *Geodesic Network*, app. F.

17. We do not distinguish in this chapter between Bell operating companies (BOCs) and the regional Bell holding companies (RBOCs or RHCs) that own them.

18. Huber, *Geodesic Network*, 3.5.

19. The interexchange carriers collected another $20 billion that went to local exchange carriers in access charges. Huber, *Geodesic Network*, 3.5.

20. The phrase "other common carrier," or OCC, refers to long-distance telephone companies other than AT&T. For example, MCI and US Sprint are OCCs. In this chapter the phrase is written out in full, despite its anachronistic

ring. Other common carriers no longer have any significant common carrier obligations.

21. The non-Bell local service companies are not affected by the consent decree and have a lower fraction of customers with equal access. GTE, however, has its own decree requiring phased-in equal access, a decree to which it consented when it acquired what is now US Sprint.

22. "Memorandum of the United States Regarding BOC Schedules for Equal Access," U.S. Dept. of Justice, CA. 82-0192, November 21, 1986. The uncertainties surrounding the NYNEX and US West numbers arise because the Department of Justice adopted a rule of thumb that if offices were not expected to provide equal access by March 1, 1988, the relevant BOC would be called upon to bear the burden of proof regarding reasonableness. Yet the data on compliance schedules do not break down the 1988 plans between pre- and post-March 1. Moreover, in the US West case, eighty-six offices had not been scheduled, and so their date of conversion was uncertain. All the numbers in the text refer to offices to which the other common carriers have requested equal access.

23. The percentage of customers is probably a little smaller; the unconverted switches are likely to rank among the largest that have not been converted (or else the other common carriers would not single them out), but they are likely to be smaller than the average converted switch.

24. AT&T pays the BOCs a "premium" access charge for its superior connections. The other common carriers claim that the premium is insufficient to compensate for the differential effects of unequal access, and, of course, AT&T has not volunteered to trade access status with one of its competitors.

25. Not surprisingly, NYNEX, which has the worst record of conversion compliance, also faces legal facilities-based competition without a block or pay requirement in the two largest states in its region, New York and Massachusetts. See Huber, *Geodesic Network,* app. B.

26. See, for example, First Boston Corp., "The Sword of Damocles; Opinion: Sell," *Equity Research Weekly Research Notes,* December 11, 1986; and Dean Witter Reynolds, Inc., "MCI Communications Corporation: Boy, Did We Get a Wrong Number," Research Note 1342, August 20, 1986.

27. Huber, *Geodesic Network,* 3.7.

28. Robert W. Baird & Co., "MCI Communications Corp.," *Technology Perspective: The Telcom Report,* August 26, 1986, p. 2.

29. In December 1986 the FCC ordered AT&T to lower its rates by 11 percent. Subsequently, MCI asked that the deregulation of AT&T be accelerated, and in August 1987 the FCC announced a proposed change in price regulation of AT&T that is the first logical step in this direction. One interpretation of MCI's stance is that it believes that AT&T would in fact price to earn transitory excess profits absent regulation, thereby providing a price umbrella for MCI. See the discussion in the text above regarding the apparent expectations of the other common carriers about AT&T pricing.

30. MCI's cash flow in 1986 cannot be accurately estimated, owing to ambiguities in the "other restructuring" expenses and an unexplained "extraordinary loss." Nevertheless, the net cash flow was at least $400 million.

31. MCI had a book loss of $448 million in 1986, owing primarily to the net effect of two extraordinary items: (a) a one-time write-down of assets and restructuring charges of $585 million and (b) a one-time gain of $104 million from the sale of assets and from antitrust settlements.

32. In addition, as their price advantage vis-à-vis AT&T has eroded, the

other common carriers have begun to stress *quality* of service in their marketing efforts; AT&T has moved recently to stress *price*.

33. Huber, *Geodesic Network,* 3.2. More than 90 percent is accounted for by MCI and US Sprint. However, two other major networks, ALC and NTN, are also growing to prominence.

34. Huber, *Geodesic Network,* 2.11.

35. See, for example, U.S. Dept. of Justice, Antitrust Division, "Report and Recommendations of the United States Concerning the Line of Business Restrictions Imposed on the Bell Operating Companies by the Modification of Final Judgment" (1987), which essentially advocates total federal withdrawal from oversight of the BOCs.

36. But see Note 39, below, regarding AT&T's planned reduction in share for certain categories of equipment.

37. The government contended at trial that the relevant geographic market for equipment sales was AT&T territory. AT&T's shares of equipment sales in the various categories in AT&T territory in 1978 were, according to the government, as follows:

Central office/switching	$2.086 b.	91%
Transmission:		
Microwave	$0.050 b.	57%
Circuit	$0.652 b.	85%
Terminal equipment:		
Station apparatus	$1.271 b.	86%
Station connections	$0.259 b.	72%
Large PBX	$0.240 b.	96%
Total	$4.558 b.	87%

The government also claimed that AT&T had the following shares of the overall telecommunications equipment business in the United States: 1974, 70.2 percent; 1975, 67.2 percent; 1976, 62.1 percent; 1977, 58.8 percent; 1978, 61.0 percent. *Source:* Plaintiffs' final statement of contentions and proof, episode 8, ¶¶ 210–213.

38. The largest independent telephone company, GTE, has its own captive equipment supplier. AT&T in the predivestiture period controlled about 85 percent of all U.S. telephone lines.

39. Certain types of equipment, such as microwave radios, were purchased in substantial quantities from outside the Bell System. This fact may illustrate the important point that a regulated monopolist's incentive to expand into competitive markets is not an absolute imperative but, rather, a response to profitable opportunities that may also impose costs. The extent of expansion is dependent upon, among other things, the relative efficiency with which the monopolist can produce goods in the competitive market. Nevertheless, while the market share of the regulated monopolist in the related competitive market may or may not reach monopoly levels, it will in general be too high from a social point of view. Despite this analysis, evidence was introduced at trial in *United States v. AT&T* that AT&T deliberately planned to reduce its market share of certain equipment types in order to forestall regulatory and legal challenges to its overall monopoly. Plaintiffs' exhibit 1351, trial transcript at 7926–7933, May 26, 1981.

40. The Centrex problem is this: Centrex provides shared central office switching services that are substitutes for CPE equipment sold by independent

PBX vendors. But Centrex is a bundled service in which both "competitive" switching and monopoly local loop services are combined. There is therefore an opportunity for the BOCs to engage in anticompetitive pricing practices with respect to the "competitive" or switching portion of the service, for the reasons discussed above. This is a particular problem when pricing of local access is already distorted by regulatory constraints.

41. Huber, *Geodesic Network,* 3.

42. *IT&T v. General Tel. & Elec. Corp.,* 369 F. Supp. 316, overturned by *IT&T v. General Tel. & Elec. Corp.,* 518 F. 2d 913, 928.

43. U.S. Dept. of Justice, *Report of the United States Concerning the Line of Business Restrictions Imposed on the Bell Operating Companies by the Modification of Final Judgment,* January 31, 1987, p. 49.

44. Obviously, these figures may reflect initial start-up costs, as well as depreciation and other accounting practices that overstate actual expenses for the purpose of minimizing tax liability, and so should not be taken too seriously. Nevertheless, regulators and antitrust authorities should regard such financial reports as signaling the continued necessity for carefully monitoring BOC ventures in competitive markets.

6

The Future Evolution of the Central Office Switch Industry

Jerry A. Hausman and Elon Kohlberg

The central office (CO) switch provides the connections among local telecommunications terminals, typically telephones and computers. CO switches in well over 95 percent of the cases provide the connection from a telephone (or computer) to another telephone or computer so that telecommunications can take place.[1] CO switches—typically referred to as Class 5 switches[2]—are best thought of as computers that have extremely sophisticated software, required by both the "real time" nature of telecommunications and the high degree of reliability established by the local telephone network and the many quite sophisticated tasks in which a CO switch engages.[3]

The CO switch market, both in the United States and internationally, is very different in 1988 from what it was in 1980. In this chapter we will examine the current situation of the CO switch market from an economic perspective. The CO switch market is a primary example of the much-discussed "globalization" of markets. Whereas in 1980 Western Electric, the manufacturing affiliate of AT&T, sold more than 80 percent of CO switches in the United States, the current situation finds AT&T with approximately half its previous market share. Northern Telecom (NTI), a manufacturing affiliate of Bell Canada, has emerged as an approximately equal competitor to AT&T. New entry by a number of European and Japanese CO switch manufacturers is occurring while AT&T and NTI are entering Far Eastern and European markets for the first time. Thus, the first

Note: David Yates has provided research assistance and the Division of Research of the Harvard Business School has provided research funds for this chapter.

major change we will discuss is the evolution of formerly national CO switch markets into international markets. Because of their extremely high development costs, modern CO switches have to be developed for international, rather than national, markets. The previous "country champion" approach, in which a government-controlled (or -regulated) telecommunications company bought almost all its CO switches from a domestic company closely affiliated with the telecommunications service provider, has eroded significantly.

The second change we will discuss is the much greater economies of scale that have arisen in the R&D of digital CO switches in comparison with the development of previous generations of CO switches. This extremely large increase in R&D has in turn increased the minimum efficient scale of production of CO switches, leading to a significant number of companies exiting from the CO switch market. Companies that previously had a significant European presence in CO switch manufacturing, such as ITT, are exiting the market.

The third major change that has occurred since 1980 is the breakup of AT&T, with divestiture of the twenty-two Bell operating companies (BOCs) that have formed seven regional holding companies (RHCs).[4] This historic change has led to increased competition throughout U.S. telecommunications markets and, in particular, in the CO switch market, in which the BOCs have aggressively developed relationships with other manufacturers besides AT&T.

After discussing the current competitive status of the CO switch market, we will turn to the possible competitive evolution of CO switch manufacturing. The current situation is defined by the consent decree entered into by AT&T and the U.S. government in 1982.[5] The Modified Final Judgment (MFJ) forbids BOC participation in the manufacturing of CO switches. The "line of business" restrictions in Section II(D) of the MFJ state that "No BOC shall, directly or through any affiliated enterprise":

1. Provide interexchange telecommunications services or information services;
2. Manufacture or provide telecommunications products or customer premises equipment (except for provision of customer premises equipment for emergency services); or
3. Provide any other product or service, except exchange telecommunications and exchange access service, that is not a natural monopoly service actually regulated by tariff.[6]

Thus, the current situation is very different from the predivestiture situation, in which Western Electric was closely integrated into an AT&T system that included the now independent BOCs. The current situation, however, may change yet again. In 1987, the government and Judge Harold Greene, the U.S. District Court judge who is in charge of the consent decree, undertook their first scheduled triennial review. Under Section VIII(C) of the MFJ, the Section II(D) "line of business" restrictions will be eliminated if "there is no substantial possibility that it [the BOC] could use its monopoly power to impede competition in the market it seeks to enter." Despite the Justice Department's recommendation that all restrictions on BOC manufacturing be eliminated, in his September 10, 1987, opinion Judge Greene declined to do so. Nevertheless, we expect that the manufacturing restriction is likely to be removed in the future.

Whether or not this change in the MFJ occurs, we foresee the evolution of the CO switch market into three to five competitive groups made up of international collaborators. The high development costs of CO switches will lead to this industry structure. The main unknown element is whether the BOCs will be able to, and will choose to, participate in the CO switch market in any role other than buyer. This unknown factor—the extent of future BOC participation in the CO switch market—has potentially important implications for the competitive future of AT&T. We think that telecommunications service competition between the BOCs and AT&T is inevitable; indeed, it has already begun to occur. Future competition will increase, however, with the adoption of advanced network features (for example, integrated services digital network [ISDN]) and AT&T and BOC provision of information services, if Judge Greene permits BOC entry into information service markets. Indeed, in his September 10, 1987, opinion, Judge Greene did permit entry of the BOCs into information services, although he restricted them to gateway and transmission services—but not content provision—of information services. No company in a competitive environment wants to be dependent on an unregulated competitor for its crucial technology. BOCs are no different, especially when the CO switch is the primary determinant of the advanced features that a BOC can offer over the local network. Thus, AT&T's position as a provider of interexchange services, information services, and CO switches may be difficult to maintain in the future. If at some point in the future Judge Greene or some other governmental body removes the restrictions on BOC participation in manufacturing, we expect that at least some BOCs will form joint ventures with switch manufacturers. AT&T, however, will be in the anomalous position of being a competitor to the BOCs, a situation that will decrease the BOCs' incentives to form a joint venture with AT&T. Thus, the current market structure, which evolved from the consent decree, does not seem viable in the longer run. Not only does this future possibility have important implications for AT&T, but the future of Bell Labs could also be in doubt. Whether a major research laboratory of the size of Bell Labs can be sustained by the future AT&T manufacturing operation needs to be examined. Alternative institutional arrangements or sources of funding could be an outcome of changes in CO switch markets.

The Central Office Switch Market Today: Globalization of Markets

The basic switching operation carries out three functions: (a) identification of the calling and called parties and completion of an electrical circuit linking them, followed by (b) maintenance of the circuit until the switch is called upon to terminate the call, followed by (c) tear-down of the call and return of the resources for further use. CO switches also perform many other functions, such as billing, that provide substantial value in addition to their basic switching role. Bell Labs' introduction of digital computers into switching provided the framework from which the current CO switches evolved. Switches now use general purpose computers as control devices and are referred to as stored program controlled (SPC) switches. The first large electronically controlled switch in the United States was the 1-ESS switch introduced by AT&T in 1965. The SPC switches represented a significant improvement over the previous electro-mechanical technology because the use of computers for control permits greater ease of modernization with new software releases and greater flexibility in use (e.g., with changes in software, a user's phone number no longer needs to be changed when he or she moves to a different location). The 1-ESS and its later development, the 1A-ESS, provide the backbone of the current U.S. telephone network. Even today, about 60 percent of the total lines served by the BOCs are controlled by these SPC analog AT&T CO switches.

Almost all new CO switches are based on digital, rather than analog, technology. That is, while both analog and digital switches are controlled by digital computers, digital CO switches process the signals as bits in a digital fashion, whereas analog CO switches process the signals as (continuous) amplitudes.[7] Digital CO switches have the advantage of treating all signals in the same manner, as streams of bits, using pulse code modulation. The time division techniques became increasingly economical as computers and switches incorporated semiconductor chips to encode signals in their binary form. The first digital switch introduced into the Bell System was the 4-ESS, a Class 4 tandem switch first installed in 1978. AT&T introduced a digital switch for toll because of the increasing use of digital techniques for interoffice transmission; however, AT&T continued to employ analog design for its CO Class 5 switches.[8] Although AT&T frequently evaluated the introduction of a digital switch for service in local central offices, each time it decided against introduction of the switch. The AT&T strategy was overwhelmed by the introduction of a Class 5 digital switch by NTI in 1979. The BOCs increasingly chose the NTI switch, in part because of technical advantages and in part because of the government antitrust suit against AT&T that was then under way. One of the key government charges against AT&T was that it had excluded competition in U.S. telecommunications equipment markets by purchasing almost all its equipment from Western Electric, the AT&T manufacturing affiliate. Thus, the Bell System felt under considerable pressure to buy equipment from vendors other than Western Electric. By 1984, when the AT&T divestiture took place, almost all new CO switches purchased were digital. NTI had by far the largest share of the U.S. market—upward of 85 percent. But by 1986, four years after the introduction of its fully digital 5E switch, AT&T had regained the lead in the new switch market. Nevertheless, both AT&T and NTI have approximately equal market shares of around 40 percent each.

NTI is the U.S. subsidiary of Northern Telecom of Canada, which holds a position in Canada similar to that of Western Electric in the predivestiture Bell System. Bell Canada owns local telephone companies; a Long Lines division; Bell Northern Research, which is the Canadian equivalent of Bell Labs; and 52 percent of Northern Telecom. Thus, the Canadian telephone system continues to resemble closely the predivestiture U.S. system.[9] In the mid-1970s, Bell Canada encouraged Northern

Telecom to develop a digital CO switch. The DMS family of switches—the DMS-10 and DMS-100—were introduced in 1977–1979. These switches are digital time division switches based on a centralized computer architecture. NTI first entered the U.S. market by selling to independent, non-Bell telephone companies. Beginning around 1982, NTI began selling to those BOCs which adopted the digital technology and employed it to meet the equal access requirements for interexchange carriers imposed by the MFJ.[10]

In 1979, AT&T began rapid development of the 5-ESS, a fully digital switch based on a distributed and modular architecture, rather than a centralized architecture as employed in the NTI switches. A distributed and modular architecture has two main advantages: the design permits the use of the same product for both small and large central offices, and it offers greater reliability than centralized systems. The 5-ESS is designed with the ability to handle the future trend to integrate voice and data and to be compatible with ISDN standards. In comparison with the AT&T 5-ESS, the NTI switch is considered to suffer from the following disadvantages: difficulty in continued expansion of installed switches to meet increased demands, difficulty in providing ISDN features, and difficulties in software maintenance. NTI's current strategy is to update the DMS series while maintaining compatibility with the existing switches and to utilize "Dynamic Network Architecture" to enhance the functioning of their product. Our research, which includes (confidential) discussions with both domestic and international telephone companies that have evaluated both NTI and AT&T digital switches,[11] leads to the finding that the AT&T 5-ESS switches have important current technological advantages over the NTI switches.[12]

Approximate market shares in the United States for suppliers of digital CO switches are given in Table 6-1. Thus, AT&T, NTI, and GTE (which sells almost exclusively to its local service company affiliates) currently have about 95 percent of the U.S. CO switch market. However, European and Japanese manufacturers, especially Siemens and NEC, respectively, are now entering the U.S. CO switch market.

Japanese manufacturers would seem an obvious factor in the U.S. CO switch market. Fujitsu, Hitachi, and NEC are all producers of large computers and are all suppliers to Nippon Tele-

Table 6-1
North American Market Shares for Digital Switch Installations
(Millions of Lines)

Manufacturer	1981	1982	1983	1984	1985	1986[a]
AT&T	—	—	.1	0.9	3.2	6.0
NTI	1.5	2.8	4.0	4.4	4.8	4.3
GTE	—	.1	.6	1.4	1.7	1.4
Stromberg-Carlson	.4	.5	.7	1.0	.2	.2
Other	.7	1.1	.4	0.3	0.2	0.1
Total	2.6	4.5	6.8	8.0	10.1	12.0

a. Sources for 1986: Huber report, Tables CO.2 and CO.4, and Northern Business Information estimates.

phone and Telegraph (NTT), which is the regulated, primary Japanese telecommunications service provider. To date, however, the Japanese manufacturers are notable by their absence.[13] Hitachi sells private switches (PBXs) in the United States, but has exited from the U.S. CO switch market after some preliminary attempts at entry.[14] NEC installed an early digital switch that it was then required to remove because of the switch's failure to meet performance standards;[15] the company is currently making a second attempt to enter the U.S. market. The absence of the Japanese from this market may arise partly because of NTT's decision to maintain an analog local network well after the initial introduction of the NTI digital switches. The Japanese manufacturers, which were the main suppliers to NTT, did not develop digital CO switches initially and fell technologically behind the rest of the world. When NTT did begin a digital progam, it decided to develop its own digital system rather than adopt the CO switches that NEC, Fujitsu, and Hitachi had each developed for the export market.[16] Northwestern Bell has ordered an NEC Class 5 switch, while Mountain Bell and Pacific Bell have ordered NEC equipment to use as adjunct digital processors for their existing analog equipment. A key component of NEC's strategy is to enter the U.S. market through the NEAX61E adjunct processors, which can be interconnected with the AT&T 1A-ESS—the primary nondigital CO switch in place through the former Bell System—to provide advanced information services such as ISDN. When these analog switches are replaced, NEC hopes to migrate the users upward to its digital CO switch, the NEAX61, which will provide Centrex service in the future.[17] Reportedly, NEC is very aggressive with respect to the price of its CO switch equipment.[18]

Many European phone systems lack features that are common in U.S. local networks—for example, itemized billing and Centrex service.[19] Thus, European manufacturers have to develop those features to compete in the United States. Moreover, the European telecommunications systems have myriad signaling standards, none of them compatible with North American standards. These disparate standards, along with a well-developed technological nationalism, have retarded joint European efforts at development of CO switches. The European telecommunications providers (PTTs) are typically government-owned monopolies that have close and protective relationships with their suppliers.[20] Until the mid-1970s each European country built up a domestic CO switch industry. Although ITT and Ericsson sold across countries, the majority of sales in each country was from the domestic producer. Many of these domestic producers initially attempted to develop a digital switch, but many were unsuccessful and withdrew from the market. Among the European companies that have withdrawn from the CO switch market are N.V. Philips (Netherlands), GEC and Standard Telephone and Cable (United Kingdom), Thomson CSF (France), and the Swiss telephone company.

ITT has been a major supplier in the international CO switch markets. ITT lagged behind other central office switch manufacturers in developing a digital switch. ITT announced its System 12 switch for the U.S. market in 1983; however, the company encountered severe software development problems with System 12 and withdrew from the U.S. market in February 1986. ITT has now merged all its equipment manufacturing operations in a joint venture controlled by the state-owned French company Cie. Générale d'Electricité (CGE).[21] The new company, Alcatel N.V., is now the second largest CO switch manufacturer in the world, after AT&T. Alcatel currently plans to sell both CGE's digital switch, the E-10, and the ITT System 12; however, the E-10 is older and less technologically advanced than the System 12, and so future development may focus on the System 12 architecture. Since the System 12 has been removed from U.S. CO switch markets, there is little possibility that Alcatel will attempt to enter the U.S. market in the near future. Instead, its operations will be concentrated in Europe and Latin America, where ITT historically has been strong.

The major remaining European companies that are attempting to enter the U.S. market are Siemens and Ericsson. Siemens has a close relationship with the German PTT and supplies approximately 55 percent of the German CO switch market. The Siemens switch is the EWSD, first introduced in 1980 but still under development. The architecture of the EWSD is a distributed switch, similar to the architecture of the AT&T 5-ESS. Siemens is making a major attempt to enter the U.S. market with a significant market share. The company has recently announced sales to Ameritech and to Bell Atlantic and is involved in ISDN demonstration projects with US West and NYNEX.

L.M. Ericsson is a Swedish company that has a significant international presence in CO switches. The AXE digital switch produced by Ericsson has been quite successful internationally, mainly in replacing ITT equipment. However, despite its success in the United States in cellular equipment and transmission equipment markets, Ericsson has not penetrated the U.S. CO switch market successfully.[22] Whether Ericsson will be able to sustain its effort in the future is unclear. Nevertheless, in April 1987 the French government chose Ericsson over both AT&T and Siemens to be the foreign partner of Cie. Générale des Constructions Telephoniques (CGT).[23] CGT is a distant second to CGE in the French CO switch market, with about 15 percent of the market, and lacks a strong technology. Thus, while the acquisition would help Ericsson's position in France, it is unlikely to have a significant effect on either U.S. or world CO switch markets.[24]

When we consider international sales of all CO switches, a strong pattern of regional sales emerges in Table 6-2. Thus, while Ericsson, Siemens, and NEC have all announced sales in the United States, they have not yet installed their switches. Nor have AT&T and NTI yet begun installing switches outside North America, although AT&T has sold switches to British Telecom and NTI has sold switches in both Great Britain and Japan. AT&T does have almost 10 percent of the "rest of the world" sales where Ericsson, with a 29 percent share, is the largest seller. Note that Japan is still not installing a large amount of new digital capacity. Despite NTI's recent sales, it

Table 6-2
International Shares of Digital Lines Placed in Service, 1986

Supplier	N. America	W. Europe	Japan	Total
AT&T	45%	0%	0%	25%
NTI	40%	0%	0%	20%
Alcatel	.3%	34%	0%	10%
GTE	12%	0%	0%	6%
ITT	.4%	12%	0%	4%
NEC	.3%	0%	24%	7%
Ericsson	0%	20%	0%	11%
Siemens	0%	6%	0%	2%
Total Lines	11.5M	5.0M	1.3M	23.1M

Source: Northern Business Information estimates. Note that these shares represent lines placed in service in 1986 and do not reflect recent sales of European CO switches to the BOCs or sales of AT&T and NTI switches outside North America.

appears that Japan will use almost solely domestic vendors— NEC, Fujitsu, Hitachi, and Oki.[25] While we expect the globalization of the CO switch market to proceed, it appears likely that Japan will remain largely a closed market, in part because NTT, the Japanese telephone company, has adopted nonconforming standards for ISDN, which could create significant entry barriers to the Japanese market.

The last important feature of the U.S. market is the presence of AT&T 1/1A analog switches, which are not scheduled to be replaced in significant quantities until the early 1990s. This embedded base of equipment arose from the predivestiture relationship of the BOCs with Western Electric. For instance, 92 percent of Pacific Bell's installed CO lines have been supplied by AT&T, and software upgrades and additional processors are supplied each year by AT&T in this market. No competition has appeared in this "upgrade" market, in part because of the complexity of the software for the switches; the primary barrier to entry into the upgrade market is maintenance responsibility if non-AT&T software is used.[26] AT&T has considerable market power here and controls the pace at which advanced features are introduced. Of course, AT&T's market power is diminished to the extent that the existing analog base can be replaced with digital switches.[27] But the installed base underlines the important economic factor that the relationship between a switch manufacturer and a buyer has an expected life of twenty to thirty years. AT&T's performance in the upgrade market has been faulted by both domestic and foreign buyers of CO switches.

Research and Development of CO Switches

The introduction of SPC switches has changed the balance of switch development from a hardware- to a software-based project. R&D for a modern digital switch is now very lengthy, with extremely high fixed (sunk) costs caused by the software development. Switch development now takes about five years before introduction, with expenditures of $100–200 million per year. During that five-year period, about 50 to 60 percent of the effort goes into hardware design. At introduction, typically only the basic call-processing features exist. Many of the features that make the switch potentially superior to previous generations exist only as promises. For the first two to three years after introduction, small systems are installed to prove the basic design. During that period and perhaps for a total of five years, development will continue at the rate of $200 million per year; however, 80 percent or more of these expenditures will be on software development. Thus, between $1 and $1.5 billion will be spent on R&D, with about 75 percent spent on software development. These R&D expenditures are approximately ten times (in constant dollars) the expenditures for an SPC analog switch. Estimates as of 1985 are given in Table 6-3.[28]

Further substantial expenditures are required to meet local telecommunications standards. European standards for signaling differ from North American standards. Bellcore, the central research organization for the BOCs, has established the Local Access and Transport Area Switching Systems Generic Require-

Table 6-3
R&D Costs as of 1985

Manufacturer	Project	R&D Costs (billions)
AT&T	5-ESS	$1.1
ITT	System 12	$1.3
Siemens	EWSD	$1.05
NEC	NEQX61	$1.0
Ericsson	AXE	$0.8
NTI	DMS-100	$0.75

ments (LSSGR) and it tests CO switches. Estimates of the development cost to meet U.S. standards for a European manufacturer are about $250 million.[29]

These extremely high development costs have transformed international CO switch markets. First, the industry has been changed from a high variable-cost industry to a high fixed-cost industry. Thus, the development costs must be funded, but they create large economies of scale. All manufacturers have concluded that they must compete in international markets; for example, a single European country is not large enough to support the development of a modern digital CO switch.[30] Second, manufacturers have decided they must compete in international markets to survive. Designing products for international markets increases development costs. Thus, the "country champion" approach is no longer viable, and significant exit has occurred from international and U.S. CO switch markets. Furthermore, doubts have been raised as to whether manufacturers that lack a large enough domestic base to finance future development of CO switches (e.g., Ericsson) will continue to be important participants in CO switch markets.

Very approximate cost estimates for North American digital CO manufacturing are presented in Table 6-4. The price per line of CO switches varies widely. The variation is largely determined by the number of lines in the CO switch, with the price per line decreasing rapidly as the number of lines increases. Nevertheless, the prices paid by the BOCs varies by only a small amount. The average price paid per line for 1987 delivery was $198, with a standard deviation of $12 per line.[31] Note that if our cost figures are accurate, little return to R&D occurs from the sale of the CO switches. Thus, return to R&D must occur

Table 6-4
North American CO Switch Manufacturing Costs

Factor	Cost Per Line
Labor	$35
Purchased chips	55
Other material	45
SGA	35
TOTAL	$170
Return to capital	$ 20

through software licensing fees and upgrades in the future, which are extremely difficult to estimate. Still, using an estimate of $1 billion for R&D and current software licensing fees, the CO switch manufacturers would need to sell some 25 to 100 million lines to recover their R&D costs. The minimum efficient scale to produce digital CO switches thus exceeds the size of a single national market, with the possible exception of the United States, Japan, and one or two European countries. Sales of 25 to 100 million lines seem unlikely for most of the current CO switch manufacturers. The total number of access lines in the United States, for instance, is approximately 120 million, with a growth rate of about 3 percent per year. Sales of digital CO switches in the United States have averaged about 10 to 12 million lines per year over the past three years. The U.S. market is about 40 percent of the current world market.[32] Thus, unless growth comes from Asian and less developed nations, these estimates would lead to the prediction of further consolidation and exit from the CO switch market. This prediction is reinforced by the expected slowdown of installation of digital CO switch lines in the United States, where equal access required by the MFJ led to a peak in demand for digital CO switches and is now largely complete. This expected slowdown in growth may well lead to increased price competition in digital CO switches, with even less return to R&D.[33] Furthermore, many fewer companies would undertake the R&D expenditure for the next generation of optical switches, given the expected required investment of two to three or an even higher multiple of the R&D costs expended to develop digital CO switches.[34] This further increase in R&D for optical switches is likely to occur because for optical switches to supplant digital switches the companies will need to offer new high-bandwidth services.[35] Since the software development costs seem highly dependent on new services such as high-bandwidth services, the expected R&D for optical switches is likely to be considerably greater than the R&D required for digital switches.

The extremely high development costs and the difficulty of gaining business across national boundaries have led to a host of joint ventures in the past three years. Both the ITT-CGE and Ericsson-CGT joint ventures have been discussed. AT&T has formed a joint venture with N.V. Philips and GTE has formed a joint venture with Siemens, both for European CO switch markets. The British comanufacturer of the System X, Plessey, has recently bought the U.S.-based manufacturer Stromberg-Carlson.[36] Northern Telecom, Ericsson, and Britain's GEC have all recently discussed further mergers or joint ventures. Thus, consolidation and reduction in the number of participants in the international CO switch market are occurring rapidly.

An unknown factor of potential importance in the near-term development of the U.S. CO switch market is the demand for digital switches from the BOCs. The rapid growth in BOC demand for these switches corresponded in large part with the implementation of equal access required by the 1982 consent decree. To provide equal access, second-generation electro-mechanical switches were replaced by digital switches.[37] However, implementation of equal access is a one-time stimulus to digital CO demand, and it is now largely complete. By the end of 1987, equal access existed for more than 80 percent of all access lines. While this fraction will rise, it will stop short of 100 percent, because BOCs are not required to provide equal access in all of their central offices.[38] Furthermore, once digital switches are deployed, their advanced features can be met with remote-switching modules that do not require replacement of the older analog switches.[39] Thus, the unknown factor is whether in the late 1980s the BOCs will find it economical to replace their SPC analog switches with digital switches. The conflicting forces here are (a) the BOC regulators' goal of maintaining low-cost basic service and avoiding large investments and (b) the future demand for ISDN-type services that would require digital CO switches. Given that the U.S. market is by far the largest CO switch market, with approximately 40 percent of lines installed in 1986, a decrease in the growth rate of BOC procurement of digital CO switches would lead to further exit and consolidation in the world CO switch market.

The economic conditions of the CO switch industry are thus characterized both by high barriers to entry and by high barriers to exit, with high-demand growth rates unlikely. Entry barriers include extremely high R&D costs, economies of scale, and the requirement of a telecommunications affiliate to provide network expertise and perhaps funding. Thus, the markets for CO switches are likely to be characterized by few participants, with more exit occurring in the future.[40] While some recent economic research has emphasized the possibility that a market with only a few participants may perform in a perfectly competitive manner, the conditions for such an outcome in the CO switch market seem unfavorable.[41] The high barriers to both entry and exit seem likely to lead to an oligopolistic outcome, with competition that might be characterized as either the "jet engine" or the "passenger airframe" market outcome. In this oligopolistic outcome, high prices and profits are not observed; rather, government subsidies and inefficient economic organization result.[42] That is, the development costs will be subsidized by the domestic country and international sales will be characterized by severe price competition because of the relatively low variable costs compared with the R&D costs.[43] Our conclusion is similar to that of the Canadian Government Inquiry on Telecommunications:

> With their domestic telecommunication market protected, manufacturers are able to recover the greater part of their development costs from domestic sales and may price their equipment for export at incremental cost. Development costs are particularly high for switching equipment. This results in highly competitive conditions in the markets that are open to competition.[44]

Exit will take place more slowly than economic factors would predict, because of countries being loathe to give up on a technologically advanced industry. Whether European cooperation will occur to the extent that it has for the Airbus is questionable, given the different standards in European telecommunications markets.[45]

Future Telecommunications Service Competition Between the BOCs and AT&T

The divestiture of the BOCs from AT&T was meant to horizontally divide U.S. telecommunications service markets into two different segments. The competitive segment was the interexchange (interLATA) markets with AT&T, MCI, Sprint, U.S. Telecom, and numerous smaller companies and resellers.[46] The monopoly-regulated segment of local exchange services was deeded to the BOCs. Furthermore, the BOCs were prohibited by Section II(D) of the MFJ from competing in interexchange, manufacturing, and information service markets so that competition between AT&T and its former partners would be minimized. However, certain areas of competition and disagreement have already appeared, along with accusations of unfair dealing, which might be expected from any divorce after seventy years of marriage.[47]

First, AT&T has approximately a 20 percent share of the U.S. PBX market. Large businesses typically choose between Centrex, a central office-based service offered by the BOCs, and a PBX. Digital Centrex on the AT&T 5-ESS switch was introduced approximately two years behind schedule. Development may have been more difficult than originally realized (as with most software projects), and equal access may have used scarce software development resources. Furthermore, the argument is made that AT&T has no incentive to endanger its $4 billion CO switch operation in favor of its $700 million PBX operation, especially given the very competitive situation in PBXs that has led to low, if any, return on investment.

Second, the BOCs complain that 800 Service Switching Point capability for 800 service database access plans was similarly delayed for the 5-ESS switch. These 800 services are a much more important source of revenue and profits to AT&T; revenue was approximately $4 billion in 1986, and little current competition exists from either MCI or US Sprint.

Lastly, the area in which the largest degree of competition would be expected is large user telecommunications services, for example, exchange access by large users to interexchange carriers points of presence (POPs). Overall, exchange access provides about 40 percent of BOC revenues. AT&T has offered Software Defined Network (SDN) and Megacom, both of which permit large users to provide their own access to AT&T POPs. AT&T has also purchased from Teleport in New York City a substantial amount of capacity that permits large users to access AT&T over the Teleport fiber optic network, rather than over New York Telephone's network. Still, the amount of AT&T competition to the BOCs for exchange access to date has been less than might be expected. Whether AT&T has not engaged in competition for large-user exchange access with the BOCs because of strategic reasons created by the MFJ is a possibility, but certainly AT&T competition has been less than economic factors alone would predict.[48] Nevertheless, Peter Huber, in his report to the Department of Justice, finds this competition between AT&T and the BOCs for large users to be inevitable:

All of the larger ICs [interexchange carriers] are aggressively moving toward vertical integration (or in AT&T's case, vertical re-integration) of their operations. All are positioning themselves to offer the end-to-end service that many large customers are still eager to purchase. Both engineering and market factors make the move toward direct connection between ICs and their largest customers inevitable.[49]

To date, the divorce has worked moderately well because of the absence of substantial competition between AT&T and the BOCs. But with the future evolution of telecommunications services, we think this situation is unlikely to continue. First, AT&T Communication's (ATTCOM) returns to "plain vanilla" interexchange services are limited by either regulation or competition. If AT&T remains the "dominant carrier" according to FCC definitions, it will likely remain subject to rate-of-return regulation. If the other interexchange carriers such as MCI and US Sprint are able to provide effective competition and regulation is removed, then the competitive process should limit AT&T's returns in what might be characterized as a "commodity" business. For either outcome, ATTCOM will seek to derive as much revenue as possible from "intelligent" services, such as "smart" 800 services that require sophisticated software and data-bases.[50] Another possible example is sophisticated audiotext services, which AT&T already provides through its "Dial It" (900 number) services. The BOCs provide similar services, in conjunction with third-party programmers, over the local network through 976 public announcement services. The value added from these services largely arises from the joint use of computer databases, together with a telephone used as a terminal.[51]

The dividing line between the monopoly-regulated sector and the competitive interexchange sector at divestiture was defined by the Local Access and Transport Area (LATA). The LATAs' definitions were derived mainly from the Census Bureau's SMSA definitions. For instance, the New York City LATA includes Long Island, New York City, Westchester County, Rockland County, and parts of Orange and Putnam counties. Under the terms of the MFJ, the BOCs are allowed to provide toll service within a LATA, that is, intraLATA service; however, they are forbidden to provide toll service between LATAs, that is, interLATA service. Nevertheless, intraLATA is an important source of revenue for the BOCs (approximately 20 percent in 1986), and an important fraction of all toll revenues (approximately 33 percent in 1986). Initially, state regulation forbade AT&T and the other interexchange carriers from competing with the BOCs in intraLATA service. By the beginning of 1987, however, thirty-six states had authorized intraLATA competition. We expect intraLATA competition to increase markedly in the near future as the interexchange carriers receive state authorizations to provide intraLATA service. This important source of BOC revenues should become an area of competition between the BOCs and AT&T, especially for large business customers.

The BOCs are currently prohibited from providing many of these services because they fall under the Section II(D) prohibition of the MFJ on information services. In 1987 and 1988, however, Judge Greene decided to grant permission to the BOCs to provide some information services. Certain information services, such as voice storage and retrieval, are not widely available to customers that do not own PBXs. Services provided to large business customers over PBXs can now be provided to small business and residential customers over public central office switches. Moreover, other information services, such as videotext and home banking, have been notably more successful abroad, especially in France, where the PTT with "Minitel" has provided the "gateway" service to link together third-party information service providers over the public network. Judge Greene's September 10, 1987, and March 7, 1988, opinions allow for BOC entry into all these information services.[52]

Advances in technology may overwhelm both regulation and the antitrust law, as applied by Judge Greene. ISDN permits the integration of voice and data traffic well beyond the current technology for all but the most sophisticated telecommunications installations.[53] ISDN will enable intelligence on customers' premises to interact in real time with intelligence in the network.[54] This development will create the potential for new services and profits. A caller to an insurance or financial services company, for example, would be identified over the signaling channel of ISDN, so that the caller's insurance policy or financial holdings would appear on the agent's terminal screen when the call is answered. Sophisticated computer databases and computer software will be required, creating a new range of

CO switch services and usage. The BOCs will want to develop both the software and the terminals for ISDN usage, and AT&T will have the same motivation. Here the competitive interaction between the BOCs and AT&T is likely to be intense. Although there has been substantial debate over whether ISDN will be successful within U.S. telephone networks, numerous ISDN trials are currently taking place, perhaps the best known of which is Ameritech's trial with the McDonald's chain.

With the potential for new services and profits, the current horizontal separation of the U.S. telecommunications network between interexchange (long-distance) and local service does not appear viable in the long run.[55] Given the rapid decrease in long-distance transmission costs created by the rapid increase in the capacity of fiber optic transmission, a high-value-added service may be alternatively provided by either an interexchange carrier or the local exchange network. Who will provide these high-value-added services will be determined to some extent by regulation and by the MFJ. Still, both the FCC, which is in charge of regulation, and the MFJ have as their announced goal the maximum amount of competition among providers. If competition is permitted, that provider which is the most able to respond to currently unclear market demands is likely to have a significant advantage. Today it takes the BOCs from three to five years to go from the conceptual stage to implementing a new service, largely because of their dependence on new CO switch software (generics).[56] Will the BOCs entrust AT&T—perhaps their primary future competitor for provision of these information services—with the task of speeding up the timetable for provision of these advanced software features, given AT&T's record in the past?[57] We find it unlikely. No competitive enterprise wants to be dependent on an unregulated competitor for an essential input, in this case the technology of the CO switch.

The Future of the CO Switch Market

The BOCs have diversified their source of supply of CO switches markedly from their previous pattern of purchasing nearly 100 percent from Western Electric. Currently, AT&T and NTI have approximately equal market shares, with European manufacturers (Siemens and Ericsson) making initial sales to the BOCs. Does this change in the BOCs' purchase patterns prove the government's contention in the divestiture trial that Western Electric's relationship with the BOCs led to an anti-competitive outcome? The government maintained that before divestiture the BOCs were paying uncompetitively high prices for Western Electric switches. The economic theory on which this claim was based held that under rate-of-return regulation, such behavior did not affect BOC profits, while it did raise the profits of AT&T overall because Western Electric was unregulated.[58] Yet the diversification of supply by the BOCs after divestiture is to be expected. A vertically integrated company, regulated or not, always does its best to buy from its in-house affiliate; a company that is not vertically integrated will seek a second or third supplier to ensure diversity of supply for both price and reliability reasons. Furthermore, at the time of divestiture the NTI digital switch appeared, and most observers (apart from Bell Labs) judged its technology as superior to that of the Western Electric analog CO switch.

We would never want to "re-try" *United States v. Western Electric,* but we bring up this part of history because of Judge Greene's decision on whether to remove the Section II(D) restriction of the MFJ to allow the BOCs to engage in manufacturing.[59] Both the Department of Justice and the FCC recommended to Judge Greene that the manufacturing prohibition be eliminated.[60] AT&T contended throughout the government trial that important efficiencies existed from vertical integration because network engineers could interact with product engineers in the design of new switches. Thus, R&D, technological innovations, and the development of new products are aided by the integration of the network provider and the switch manufacturer. This vertical integration exists in virtually every foreign country, including Canada.[61] Peter Huber, in his report to the government on the MFJ restrictions, concludes, "[T]here are very real and substantial economies in manufacturing equipment in partnership with LEC (local exchange company) affiliates."[62] Nevertheless, in his September 10 opinion Judge Greene declined to remove the MFJ restriction on manufacturing by the BOCs, and so, at least in the near future, BOC entry into the manufacture of CO switches will not occur.

A further economic question is whether a BOC and an unaffiliated manufacturer together have the same incentives for development of new products as a BOC and an affiliated manufacturer do. The MFJ prohibition on manufacturing appears to create different incentives. The MFJ limits the returns to a BOC innovation to only its network because a BOC would not be allowed to take an equity interest in a joint venture with a manufacturing operation. Also, depending on the type of innovation, the BOC's return might be limited by rate-of-return regulation that decreases incentives to innovate. Lastly, a manufacturer will have greater incentives to innovate if it has guaranteed access to markets, in this case the network. All these factors would appear to support vertical integration of CO switch manufacturers with network service providers.[63]

If the manufacturing prohibition of the MFJ is eliminated at some future date, we would expect some of the BOCs to form joint ventures with CO switch manufacturers, rather than forming their own manufacturing affiliates. Given the extremely high R&D costs and the associated economies of scale, a regional holding company (RHC) could not economically develop and manufacture CO switches in the predivestiture Western Electric pattern of selling only to its affiliated BOCs. A single RHC market is not large enough to support a "captive" CO switch-manufacturing affiliate. The consolidation of European manufacturers has made this economic factor quite evident.[64] Nor would regulators permit an RHC to buy only from a manufacturing affiliate that made no outside sales. Thus, we would expect BOCs to form joint ventures with existing CO switch manufacturers, while still maintaining a diversity of supply for both economic and regulatory reasons.

If the BOCs are permitted to form joint ventures, will they form joint ventures with AT&T or only with foreign manufacturers? Clearly, the antitrust laws would limit the number of BOCs (RHCs) that could form joint ventures with AT&T. But it is not unlikely that the government would permit two or three RHCs to form joint ventures with AT&T were the MFJ restriction on manufacturing eliminated. Indeed, Huber, in his report to the government, states that "it would be a national disaster of major proportions" if the RHCs were not permitted to form alliances with AT&T.[65] He bases this judgment on the enormous R&D requirements to develop the next generation of optical switches and the importance of Bell Laboratories to the United States. While we agree with Huber that the RHCs should be permitted to form joint ventures with AT&T, we are not confident that the RHCs have the economic incentives to form such joint ventures with a current and future competitor, given the corporate structure of AT&T that resulted from the divestiture.

However, even at present, when Judge Greene has decided to continue the MFJ prohibition on BOC involvement in the manufacture of telecommunications equipment, the competitive rivalry between the BOCs and ATTCOM will lead the BOCs to reduce their dependence on AT&T and turn increasingly to other CO switch manufacturers, such as Siemens and perhaps NEC. The BOCs will not want to be in the position of technological dependence on a competitor, nor will they want to discuss further service plans with the manufacturing affiliate of a competitor.[66] If the future structure of the CO switch market includes AT&T, Siemens, NEC, and perhaps NTI and one or two additional European or Japanese competitors, then the only U.S. manufacturer will be at a long-term competitive disadvantage for the major source of U.S. demand for CO switches.

A possible way around this problem would be for AT&T to separate AT&T Technologies, the manufacturing arm of AT&T, from ATTCOM, the long-distance arm of AT&T. A manufacturing operation, treated as a separate subsidiary, would then be able to treat ATTCOM and the BOCs on the same basis. In the short run the economies from vertical integration might be reduced, but these economies would be replaced by interaction between AT&T Network Systems and the BOCs, especially if the BOCs are permitted to enter into manufacturing, which would be likely to lead to joint ventures between some of the BOCs and AT&T. Although some might claim that such a result would "undo" the divestiture of the BOCs by AT&T and lead to the former problems, such as a claim would miss the crucial point that seven RHCs now exist that would purchase CO switches from a variety of manufacturers. These purchases would provide the regulatory benchmarks lacking in the predivestiture relationship between Western Electric and the BOCs.[67]

In the predivestiture era, Bell Labs was financed by a "tax" on local telephone service. Since more than 90 percent of all households and all businesses had telephones with a very low price elasticity of both access and usage, presumably the economic distortion created by the tax was not especially large. Indeed, it may well have been less than funds raised by other types of government taxes. However, Bell Labs is now funded by AT&T revenues. If our future scenario is close to correct, the future of Bell Labs may be in doubt, an outcome that could create future difficulties, since the development costs of the next generation of optical switches is expected to be on the order of $2 billion or more. To the extent that Bell Labs is seen as an important national resource for advanced research, its future may be more important than the profitability of AT&T Technologies. The current and future competitive situation resulting from the AT&T divestiture could create important problems for Bell Labs that should not be lost sight of. The current technological superiority of the AT&T CO switch is widely recognized among much of the telecommunications industry, and this superiority in U.S. switching technology should not be lost because of regulatory and antitrust solutions to problems that occurred mainly in the interexchange markets between ten and twenty years ago.

Conclusion

The technological change in CO switches that occurred with the move from analog to digital switches has led to significant changes in international CO switch markets. The most important technological feature of digital CO switch development is the very large R&D costs. These costs are large both in absolute terms—$1 to $2 billion—and in relative terms when compared with the manufacturing costs of a CO switch. Another important feature of CO switch development is the lack of a patent at the end of the development process. In the chemical and pharmaceutical industries, where R&D are also quite large, the existence of a patent increases the prospect for recovery of R&D expenditure when manufacture of the product begins. The two final features we emphasize in CO switches are that (a) except in the United States, purchase decisions are made by a centralized authority, either a PTT or a regulated company, and even in the United States, buyers are significant in relation to the overall world market and can exert significant pressure on sellers, and (b) political considerations of national defense and national pride for a technologically advanced industry assume great importance.

The combination of these factors will lead to there being probably three but no more than five long-term competitors in the

CO switch industry. North America will have AT&T, a European industry will center on either Siemens or Alcatel (CGE), and we expect a Japanese industry to center on NEC. NTI and Ericsson may also remain important competitors, or perhaps another Japanese manufacturer will become a significant force in the market, although we consider it unlikely. This expected outcome of three to five competitors will be somewhat slow to happen because of the political considerations we have noted. Nevertheless, we see little likelihood of there being new entrants into the CO switch industry, and exit and consolidations seem to be occurring at an increasingly faster rate.

While AT&T has responded well so far to competition in the U.S. digital CO switch market by recapturing a significant amount of market share that it had lost to NTI, we believe that the most important problems in international competition for AT&T are just now beginning. With the market structure described above, we believe the competitive outcome will be one in which the R&D expenditures are subsidized by domestic countries or by international consortia, such as now fund the Airbus. The sale prices of the CO switches, however, will be much closer to marginal costs of production than to average cost, which would include R&D expenses. Additional revenue from software licenses and updates is unlikely to be sufficient to cover R&D expenditures. R&D will continue to be funded mainly on political grounds, with national security and technological advancement continually put forward. In this situation, the legal framework of a market structure that assumes no significant competition between AT&T and the BOCs, a concept that was embedded in the MFJ at the insistence of the Department of Justice in 1982, is very dangerous for the future of the U.S. CO switch industry. Future international competitive realities must be considered in predicting the evolution of CO switch markets.

Notes

1. Large users may avoid the use of CO switches altogether. They may have their own private network in place, controlled by private switches (PBXs), or they may purchase unswitched transmission capacity from the local exchange carrier or another supplier. However, these alternative forms of transport are typically used for interexchange traffic.

2. The numbering system on CO switches is a holdover from the designa-

tions adopted in the predivestiture AT&T system. The other major type of switch is the Class 4, or tandem switch, which basically connects Class 5 switches together. Most of this chapter will be concerned with Class 5 CO switches, which make up more than 90 percent of U.S. CO switch sales.

3. The AT&T Class 5 switch is engineered to the criterion of two hours of downtime in forty years.

4. In other countries, deregulation and privatization of the telecommunications service providers are occurring. In both Great Britain and Japan the formerly government-owned monopoly providers of telecommunications have now been privatized. In Great Britain, British Telecom has begun purchasing central office switches from North American and other European manufacturers. These foreign purchases are a distinct change from the previous situation of all domestic CO switch purchases.

5. *United States v. Western Electric Co., Inc, et al.; Notice of Entry of Final Judgment,* 47 Fed. Reg. 40392 (September 13, 1982).

6. Section VIII(A) of the MFJ permits the BOCs to provide, but not manufacture, customer premise equipment.

7. Since amplitudes have a continuous representation while bits have a 0–1 representation, the tolerances in analog transmission and switching exceed the tolerances required in digital transmission and switching.

8. Today all new interoffice (trunk) connections are fiber optic, which employs a digital technology. However, almost all telephones are still analog devices. Whether digital CO switches are more economical than analog CO switches has never been definitely resolved. Market developments have made the debate irrelevant by now.

9. Bell Canada buys only Northern Telecom switches, a situation similar to the predivestiture situation in the United States, wherein the BOCs bought almost all their switches from Western Electric. The Canadian government called for the divestiture of Northern Telecom by Bell Canada in the early 1980s, but the recommendation was later dropped. See *Telecommunication in Canada, Part III: The Impact of Vertical Integration on the Equipment Industry,* a report by the Canadian Restrictive Trade Practices Committee, 1983. The conclusion of the committee was: "The evidence in this inquiry does not establish that, on balance, the separation of Bell and Northern would improve performance in the telecommunication equipment industry or in the delivery of telecommunication services by Bell and other carriers" (p. 207). However, only about 28 percent of Northern Telecom's total switch sales are to Bell Canada, while Western Electric made very few sales outside the Bell System before divestiture. Thus, regulation of Bell Canada has "benchmark" sales of Northern Telecom for comparison purposes, which predivestiture regulators of AT&T lacked.

10. NTI does most of the manufacturing of digital CO switches sold in the United States at U.S. locations. In 1983, Northern Telecom reported to the Canadian government that it did 50 percent of its R&D at U.S. locations.

11. These conversations took place in 1986. The relative rankings could well change rapidly as new modifications to switches are developed and introduced.

12. GTE is the other major domestic manufacturer of CO switches, which it primarily sells to its affiliated local telephone companies. GTE recently sold its non-U.S. switch operations.

13. NEC claims to have the second largest amount of lines in place and orders for digital CO switches in export markets, trailing only Siemens.

14. Hitachi is reported to have decided against entering the U.S. market because of the required twenty-year commitment to maintain and upgrade the switches. The future of Hitachi, Fujitsu, and Oki, the primary Japanese switch

manufacturers in addition to NEC, in world markets may be limited by the nonconforming standards for advanced information services and ISDN adopted by NTT, the Japanese telecommunications company.

15. The NEC switch was installed by Rochester Telephone, an independent telephone company. After the switch was removed, NEC redesigned it. Nevertheless, NEC has continued to have credibility problems in selling to U.S. telephone companies.

16. Even now, when NTT has decided to accelerate the introduction of digital CO switches, the software developed for the NTT D70 cannot be used in the export market. It is very unlikely that the D70, developed for NTT by NEC and the other Japanese telecommunications manufacturers, will become a significant product in international digital CO markets.

17. NEC introduced a small subset of Centrex in September 1987. Further development of Centrex is planned for its digital switch. Centrex, while used extensively in the United States, is little used in either European countries or Japan. The software development costs to include Centrex in a digital CO switch are quite large. NEC executives indicated that they believe software development in export markets so that specific customer needs can be met is the key component of future export competition.

18. NEC executives stated that they do not believe they are large enough to become the "third supplier" to the entire U.S. market, together with AT&T and NTI. Instead, they aim to establish relationships with either two or three of the seven RHCs.

19. For instance, System X, the digital CO switch developed by GEC and Plessey in Great Britain, initially lacked both features. British Telecom is buying AT&T digital CO switches to provide Centrex service in London. No plans currently exist to develop System X for the U.S. market, in part because of the high development costs to add these features.

20. The major exception is Great Britain, where British Telecom has been privatized and a second competitive supplier primarily of long-distance service, Mercury, has been permitted to enter the market. British Telecom has also begun purchasing about 20 percent of its central office switches from Ericsson, despite the availability of System X, a digital switch produced by GEC and Plessey, both of which are British telecommunications equipment companies. British Telecom also plans to purchase AT&T switches to provide toll-free, 800-type service and Centrex service in London.

21. CGE, which controls the joint venture, was scheduled to be privatized in 1987.

22. Very recently Ericsson did announce sales of digital CO switches to US West (*Communications Daily*, April 8, 1987).

23. AT&T has formed a joint venture with N.V. Philips for CO switch sales in Europe. Except for sales to British Telecom, AT&T/Philips has been unsuccessful so far in European CO switch markets.

24. Political considerations seem important in the French government decision on CGT. AT&T/Philips and Siemens were the two main contending suitors for CGT. Unwilling to decide on either the Europeans (the European Community) or the Americans over each other, France decided to pursue a "neutral course," according to unnamed French officials (*New York Times*, April 30, 1987, p. D2).

25. Fujitsu placed 40 percent, Hitachi placed 19 percent, and Oki placed 17 percent of the digital lines in Japan in 1986, according to Northern Business Information estimates. Since its recent privatization, NTT has announced a more rapid installation program of digital CO switches.

26. The MFJ permits the BOCs to create and sell software for CO switches.

27. The rate at which the BOCs will replace their analog switches with digital switches is not altogether clear. Digital switches can be placed in strategic positions to limit the requirements for replacement of analog switches unless considerable demand exists for the advanced features of digital switches.

28. An AT&T executive estimated that as of 1987 these R&D costs might be multiplied by two times to achieve a more accurate estimate.

29. These costs may decrease in the future, when initial development of a switch embodies the capability to meet foreign requirements.

30. Nor would a single RHC be large enough to support development if it sold only to its affiliated BOCs.

31. Source: Submission by the BOCs to Peter Huber. See Huber, *The Geodesic Network: 1987 Report on Competition in the Telephone Industry* (Washington, D.C.: U.S. Department of Justice, 1987), Table CO.12, p. 14.19.

32. Furthermore, it is far from certain that advanced analog CO switches in the United States will be replaced by digital switches.

33. In an industry that lacks important considerations of national pride or national defense, the number of competitors would be importantly influenced by considerations of minimum efficient scale. In an industry such as CO switches, national considerations lead governments to subsidize their domestic companies. The number of worldwide competitors may then be either larger, with small countries subsidizing their manufacturer, or smaller, because the need to obtain market share leads to price competition among the large manufacturers, which makes the small manufacturers' losses too large to be subsidized. Before digital switches, the former situation prevailed; the increase in minimum efficient scale caused a "tipping" from the previously stable situation of too many competitors to the current situation.

34. Lower labor costs for non-U.S. manufacturers would affect these calculations. However, much of the assembly of digital CO switches is already highly mechanized, with a high use of robot assembly by both AT&T and NTI. Our estimates indicate that only about 20 percent of the manufacturing cost currently goes to labor costs.

35. Note that universal agreement does not exist over whether the next generation of optical switches will come into use. A senior executive of Bell Laboratories claimed that another generation of switches will not occur. However, almost all other scientists and engineers with whom we discussed this issue do foresee the development of optical switches. Furthermore, Bell Labs' foresight into the future developments of central office switches has not been outstanding, given its decision in the 1970s that the current digital generation of central office switches was not economical.

36. In 1987, Plessey/Stromberg-Carlson received an order for 600,000 lines for digital CO switches for delivery in 1988 and 1989 from South Central Bell (*Telephony,* April 13, 1987). This order may present significantly better prospects for Stromberg-Carlson, which many analysts expected to exit the market before its purchase by Plessey.

37. Equal access could be implemented straightforwardly on the SPC analog switches because of their software control.

38. Equal access is expected for about 89 percent of all subscriber lines by 1989.

39. Fiber optic technology makes this remote placement strategy increasingly economical.

40. This outcome is unlikely to depend on whether the BOCs are allowed to enter the manufacturing market.

41. See W. Baumol, J. Panzar, and R. Willig, *Contestable Markets and the*

Theory of Industry Structure (New York: Harcourt Brace Jovanovich, 1982). "Perfect contestability" requires no entry barriers, no exit barriers, and the ability of the entrant to capture the market before the incumbent(s) can respond (i.e., "hit and run entry"). None of these conditions are satisfied in the CO switch market. Further exit barriers arise because of the obligation to maintain a switch for twenty to thirty years after installation.

42. While the presence of only a few firms in an industry always raises the potential problem of price-setting "collusion," we expect the technology and the structure of the industry to lead to prices being below average cost (including R&D) for most of the time.

43. Given the requirements of the technology with the likely outcome of only a small number of firms producing an essential input (e.g., either CO switches or airplanes) specific to an activity carried out by only a few firms, economic theory would lead to the prediction of extensive vertical integration to avoid small-number bargaining problems, such as opportunism and bilateral monopoly (or oligopoly) problems. See O.E. Williamson, *Markets and Hierarchies* (New York: Free Press, 1975). Thus, vertical integration reduces conflicts of interest that contract relationships among independent firms are typically incapable of doing as well. Vertical integration is considered in a simplified model of international trade by E. Helpman and P. Krugman, *Market Structure and Foreign Trade* (Cambridge, Mass.: MIT Press, 1985, Chap. 13). However, because of the nationalism inherent in national airlines and in national telecommunications companies, less vertical integration across countries occurs than would otherwise be expected. Thus, the extent of multinational companies is limited partly by political considerations. This outcome is likely to be less efficient than an outcome with greater integration, but political considerations of nationalism seemingly outweigh more refined economic calculations here.

44. Canadian Restrictive Trade Practices Committee, *Telecommunications in Canada*, p. 199; footnote omitted.

45. The "Gang of Four"—Plessey (U.K.), Siemens, CGE-Alcatel (France, also known as CIT-Alcatel), and Italtel—have agreed to pool R&D resources. It is unclear what degree of joint effort will arise from this combination.

46. How competitive interexchange markets will be is subject to question and economic analysis. While the government (Justice Department report to Judge Greene) claims a high degree of competition, the loss of the 55 percent advantage in access costs to AT&T's competitors has led to large (accounting) losses of approximately $500 million to both MCI and US Sprint, which is the combination of the formerly separate competitors GTE Sprint and U.S. Telecom. Equal access elections are now largely over, and AT&T has either maintained or even increased its market share. Especially troublesome is the stock market reaction to AT&T and MCI's combined call for deregulation of AT&T by the FCC in March 1987. MCI's stock rose by 40 percent which leads to the economic inference that deregulation might lead to a significant increase in prices. Unless one takes the view that FCC regulation is holding down AT&T's prices to subcompetitive levels, MCI's ability to compete on a "level playing field" may be questioned. However, neither AT&T nor the FCC can afford to let MCI and US Sprint disappear. But the amount of effective competition for residential and small business customers in the future is uncertain. Future developments, such as the completion of pending fiber optic networks, will increase large business competition but is unlikely to have a significant effect for small businesses or residential customers.

47. Indeed, relationships have been deteriorating lately, with the govern-

ment's recommendations to Judge Greene that the BOCs be allowed to enter manufacturing and information service markets. One RHC, US West, has claimed that it is not subject to the consent decree which it argues was a deal between AT&T and the government, but this legal argument has been rejected to date.

48. Under Section VIII(C) of the MFJ, BOCs are permitted to enter markets forbidden to them under Section II(D) if there is "no substantial possibility that they can impede competition in the markets they seek to enter." If large user exchange access were deemed a competitive market, the BOCs could receive permission to provide interexchange service to large users, which comprise one-third or more of the business interexchange toll market, depending on the definition of large users. The prospect of BOC entry into interexchange markets in competition with AT&T could radically alter the current competitive situation. For instance, California is divided into twelve LATAs, among which Pacific Bell is not permitted to provide interexchange service. Pacific Bell entry into California interexchange markets for large users would likely occur if permitted by a waiver of the MFJ under Section VIII(C).

49. Huber, *The Geodesic Network*.

50. Note that the BOCs are now developing the capability to offer certain features of 800 services that they are currently permitted to offer under the terms of the MFJ.

51. Ameritech estimates that these services generated more than $500 million in revenues in 1986, with approximately 90 percent accruing to the BOC 976 services (Ameritech submission to Huber, August 11, 1986).

52. Both the Department of Justice and the FCC recommended to Judge Greene that all information service restrictions on the BOCs be eliminated. They determined that market forces and regulation decrease the potential anticompetitive problems to the extent that the potential benefits from BOC participation in information service markets outweigh the potential problems, and elimination of the complete information service restriction of the MFJ is called for. Judge Greene rejected this recommendation and removed the information services restriction only partially, forbidding the BOCs to engage in content provision (e.g., electronic Yellow Pages) for information services.

53. The potential importance of ISDN is demonstrated by Siemens' U.S. market entry strategy, which is largely based on ISDN features for its CO switch.

54. Huber, in his report to the government on the current state of competition in telecommunications, has emphasized the increase in intelligence on customer premises (*The Geodesic Network*). Huber's viewpoint has been challenged by AT&T and other commentators as being incorrect.

55. We also expect competition to develop between AT&T and the BOCs over large customers in other areas, such as private networks. However, some forms of this competition will continue to be prohibited by the MFJ, at least over the next three years.

56. Some analysts claim that open network architecture (ONA) may eliminate this competitive dependence because of implementation of future technology into current CO switches through "plug-in boards." However, this claim overlooks the important fact that a standard operating system does not exist for CO switches. Indeed, proprietary operating systems for CO switches are closed, and we expect this situation to continue. It is extremely unlikely that a standard software system will evolve for CO switches. Thus, plug-in boards will be severely hindered or even eliminated by the proprietary software, since the various application programs are interleaved with the main operating

system in a much more complicated manner than with mainframe computers. Adjunct processors are unable to provide many features (such as advanced Centrex) and require software unique for each CO switch for the features they do provide. Thus, these features will continue to be provided by the CO switch, and AT&T controls the software for its 5-ESS digital switch. Furthermore, since we are told that the operating system for this switch is considerably more complicated than that for an IBM mainframe computer, we find it extremely unlikely that a BOC or third party will modify the software in a 5-ESS digital switch to provide advanced features in competition with AT&T.

57. Some analysts claim that AT&T originally lost its competitive advantage in PBXs partly because of its failure in the 1970s to provide least-cost routing for long-distance calls. Here least-cost routing would have created additional competition for AT&T Long Lines division. Western Electric was slow in developing this capability for its PBXs, but PBXs from NTI and Mitel had other advantages over the AT&T PBXs.

58. This theory was never proved during the trial, and Judge Greene rejected it in his decision on AT&T's Motion for Summary Judgment.

59. The *United States v. Western Electric* is "re-tried" in the book by Steve Coll, *The Deal of the Century* (New York: Atheneum, 1986).

60. Unsurprisingly, AT&T opposed the government's recommendation.

61. Northern Telecom made similar claims to the Canadian Restrictive Trade Practices Commission when the divestiture of Bell Canada and Northern Telecom was under consideration. The Canadian government report states: "The argument was advanced that vertical integration between Bell and Northern enables all parties to draw freely on the total knowledge-base and removes the need to either withhold information or enter into detailed contracts regarding its use. . . . This [lack of integration] could seriously inhibit new product development in areas such as switching." (*Telecommunications in Canada*, p. 89). The government's recommendation to divest Northern Telecom from Bell Canada was subsequently withdrawn.

62. Huber, *The Geodesic Network*, p. 14.13.

63. This reasoning leads to the question of why AT&T was not divested vertically, with perhaps three RHCs created, each with its own manufacturing capability. Participants on both the government side and the AT&T side have told us that this possibility was never considered. A vertical divestiture would have been in conflict with both the government's economic theory and AT&T's overoptimistic assessment of the competitive ability of Bell Labs and Western Electric to prosper in a nonregulated environment.

64. The large RHC purchaser of CO switch lines in 1985 was Bell Atlantic, which put into service about 1.4 million lines of new switching capacity (Huber, *The Geodesic Network,* p. 14.13). This purchase represents considerably less than 10 percent of the world market, which is too small to support a CO switch manufacturer. AT&T and NTI each produces in excess of 5 million lines per year.

65. Huber, *The Geodesic Network*, p. 14.25.

66. Of course, CO switch manufacturers (e.g., NTI) compete with the RBOCs in areas such as PBX marketing. These areas of competition do not include telecommunications service provision, which we foresee to be the main area of BOC-AT&T competition.

67. Beside direct benchmarks from purchases of CO switches, the presence of independent BOCs makes "yardstick" regulation possible, even in the presence of differences across the BOCs. See A. Shleifer, "A Theory of Yardstick Competition," *Rand Journal of Economics* 16 (1985), p. 319.

Part Three
The Special Needs of Large Users

7

The Role of the Large Corporation in the Communications Market

James L. McKenney and H. Edward Nyce

Summary

History, economics, and corporate strategy ensure that the large corporation is, and will continue to be, an active participant in the telecommunications market. The ability to develop and manage its own communications services provides the global business with quick links to customers and better business operations through reliable, adaptive communications channels. With the advent of distributed computing in the late sixties, major corporations came to depend upon internally managed data communications services. While at the time most corporations viewed data communications management as an increased cost, they also believed that it was necessary for strategic reasons. With experience in managing communications, they broadened their communications services to include office support as well as computer services. Today the volume of internal demand and the productivity of new technologies provide favorable returns on investments in additional communications capacity. This trend seems to be increasing, and corporate managed telecommunications (CMT) is a growing phenomenon.

Two forces—the usefulness and growth of data communications and the globalization of competition[1]—combined to give a strategic thrust to the use of data communications in the late 1960s. The ability to generate and distribute computer-generated information created the opportunity to provide more effective management information quickly to disparate geographical areas. This new communications service grew to be a vital enabling force in the globalization of industry and allowed firms to link factories to markets worldwide.

Growth of the Corporate
Telecommunications Market

The expansion of firms into new global markets in the 1960s was the first strong impetus to the active management of communications. The need to coordinate the distribution of products to distant markets placed a premium on having communications links between countries and in most instances within countries. The lack of reliable international telecommunications, coupled with two-year bureaucratic delays to change service, led most large firms to develop their own leased-line private networks.[2] The firms formed organizations to manage and maintain these communications systems. The early systems were primarily telex; although relatively slow, they were the only alternative available. As new digital technologies appeared, the existing telex systems were replaced by computer-based communications systems. By 1977, 75 percent of international data flow was on leased private line,[3] the majority between computer and terminals.

The primary force that stimulated the growth in CMT was the continuous improvement in microchip technology that produced more accurate and reliable communications at significantly lower costs. The advent of Very Large-Scale Integration (VLSI) technology allowed the development of communications switches that could link high-speed computers together to exchange data at microsecond speeds. This advancement led to a range of more efficient use of resources—for example, in inventory control. Many of these developments subsequently became the basis of other innovations along the corporate value chain, such as electronic inventory systems and new distribution systems at American Hospital Supply Corporation.[4] These innovations in turn changed the nature of computer-based information from a mechanism facilitating management to an active means of competition. In the transition, telecommunications moved from a necessary service function to a strategic enabling resource with significant bottom-line impact. The result is that telecommunications services are now an integral component of market strategies. For several multinational enterprises (MNEs), telecommunications is the second largest purchased expense after staff.[5]

As businesses reached out to different markets and sources of supply, they linked these entities to existing locations through telecommunications. Computers were linked to each other and to terminals that could display information and receive data as well as connect to remote control devices. Distributing information electronically enabled companies to gather and disperse information independent of geography. The geographic growth was compounded by the shift to on-line entry for data capture for computers and increased computer-to-computer communications. Its dramatic cost increase and complex demands forced managers of large computer centers to seek more efficient means of data communications.

The new data links required more accurate transmission of the signal and more flexible rates of transmission than those afforded by voice service. The regulated telephone companies responded slowly to this growing demand. Ninety years of honing a system in a twenty-eight-year development cycle to deliver voice service did not equip AT&T to respond to these needs in the time frames required.[b] Nor did the AT&T culture provide much room for suggestions on the definition of the service—that is, on what was needed for data communications. Instead, the standard response was modems on voice lines; their speed was slow, and their requirements were cumbersome.

Encouraged by the Advanced Research Program Agency (ARPA) of the Defense Department's development of AR-PANET and by such corporate leaders as Ford, General Electric, IBM, and others, several companies began to develop their own data networks from 1965 to 1970.[7] In the process, they learned how to manage data communications systems by developing software control systems for the transfer of data on leased lines. They also discovered how to reduce costs by investing in switching systems that enabled them to lease low-unit-cost, large-volume transmission lines and thus to distribute services internally at significantly lower costs. By the late 1970s, several MNEs managed worldwide data communications services.[8]

The expansion of office automation in the mid-seventies was a parallel development. Because of organizational traditions and "benign neglect" of office design, this effort often started independently of the data processing (DP) group that managed data communications. An apprehension that the involvement of data processing would lead to large, complex, centralized systems contributed to this independent development.[9] Early efforts were focused on linking word processors by electronic mail, which often led to wiring the office for in-house electronic communications to improve information exchange. Following the decision to allow purchase of customer premise equipment (CPE), office managers purchased private branch exchanges (PBXs) to reduce voice service costs and linked them to the internal wiring. These newly automated offices expanded their means of providing communications services by fashioning complex private line leasing arrangements to reduce costs and, in general, by pursuing a rich gambit of make-or-buy decisions to provide voice communications services.

The growth of terminals linked together to support office work created situations in which multiple networks were serving the same area. One firm found seven different leased-line networks in an office; combining the lines into one through the use of concentrators and switching equipment saved $40,000 per month.[10] The increased awareness of office communications costs, coupled with new switching technology, quickly led to the internal design of local distribution systems and the development of general purpose office communications systems. In the early stages of office systems development, most organizations maintained separate departments to manage communications. As the office effort grew, the totality of communications costs became more apparent, and most companies have now consolidated data communications into a single organization.[11]

Thus, the period from the 1960s to the present reveals an evident trend toward CMT, with numerous factors converging to produce that trend. A case in point is Manufacturers Hanover Corporation, the nation's fifth largest financial institution, which has developed and at present manages a proprietary worldwide telecommunications network that links to customers and to other financial institutions and that provides all internal communications services. In the early 1970s, data communications was an integral part of DP and perceived by Manufacturers Hanover as a necessary function that had to be developed and supported. Increasing globalization of the financial markets; the obvious need for accuracy, reliability, and security in financial transactions; and a growing reliance on office automation to carry out the myriad details of operations and service were all at work to affect the development of data communications and to shift the perception of its role within the institution.

Manufacturers Hanover's experience was typical of how MNEs grew to provide CMT.

Corporate Telecommunications at Manufacturers Hanover

Overview

Corporate telecommunications per se did not exist at Manufacturers Hanover (MH) before 1976. The jobs now done by most of the institution's telecommunications staff had not yet even been thought of. Rather, bank staff grew into their jobs as corporate telecommunications developed—in volume, in geographical scope, in the uses to which it was put, and in the growing importance it assumed for the bank's method of doing business.

Today, more than a decade later, MH has an internal staff of 270 in the telecommunications group. A proprietary and global communications network links all of the institution's disparate and far-flung offices to huge databases of information about customers and their financial transactions; a map of the network is shown in Exhibit 7-1. The total direct expense for telecommunications during the single year of 1986 was $96.5 million. Voice accounted for approximately 60 percent of the total, and data accounted for 40 percent. All purchased telecommunications services, lines, and usage amounted to $69.3 million.

The prime data network, GEONET, consists of 34 nodes, 484 concentrators, and 975 dial-in ports. More than 8,000 terminals are directly connected to it, and several thousand additional terminals are indirectly connected to it by shared dial-in ports. All the bank's major data centers connect to the network, and connectivity extends out to departmental machines and end-user personal computers within the organization. The network is maintained around the clock and around the world. It connects 63 cities and covers 32 countries and is key to the global operations capability of MH. John J. Evans, a vice chairman at Manufacturers Hanover, and the man largely responsible for the institution's early commitment to telecommunications, has said that "the manner in which we now gather, assemble, and disseminate information has transformed forever the way the business of banking is conducted."[12]

One barometer of that change is the fact that, every day, more than $1 trillion is transmitted around the world over com-

Exhibit 7-1
GEONET 1987 International Network

SYDNEY

TOKYO
OSAKA
SEOUL
SHANGHAI
TAIPEI
HONG KONG
MANILA
KUALA LUMPUR
SINGAPORE
BEIJING
BANGKOK
JAKARTA

BOMBAY

CAIRO
BAHRAIN

ISTANBUL
BUCHAREST
AMSTERDAM
HAMBURG
HANOVER
STUTTGART
DUSSELDORF
FRANKFURT
MUNICH
PARIS
ZURICH
ROME
MILAN
STOCKHOLM
OSLO
LONDON
GENEVA
LUXEMBOURG
BERNE
GUERNSEY
MADRID
LISBON
BARCELONA

CLEVELAND
BOSTON
BUFFALO
LATHAM
ROCHESTER
ALBANY
HICKSVILLE, N.Y.
NEW YORK CITY
TEANECK
LIVINGSTON
HUNTINGDON VALLEY
CHARLOTTE
WILMINGTON
DANVILLE
ATLANTA
MIAMI
CARACAS

CHICAGO
DENVER
SAN FRANCISCO
LOS ANGELES
DALLAS
HOUSTON
BOGOTA
BUENOS AIRES

Source: Manufacturers Hanover Corporation.

puterized networks. The phrase *banking hours* has no meaning anymore. It is always banking hours—not only for financiers and corporate customers whose areas of concern stretch from Tokyo to New York to London to Singapore, but also for the retail customer who jogs down to the automated teller machine (ATM) at midnight to obtain cash, make payments, or get the latest information on his or her checking account balance.

But it is not only the way banks do business that has been affected; it is the nature of the business itself. Information about money and about money movements has become almost as important as money itself. How well, how fast, and how cost-effectively a bank gathers and transmits information have become criteria of service quality—and not just for cash management services. The chairman of Manufacturers Hanover Corporation, J.F. McGillicuddy, talked about the new realities of a bank's credit function:

> The global information systems we bankers have put in place are now beginning to alter the very process of credit itself. Our ability to deliver information on demand worldwide is giving new meaning to such trends as the securitization of credit assets and the disintermediation of the lending function. These basic trends of course have been with us for some time. But how they ultimately play out will depend in no small way on the technology that now allows us to communicate information quickly, and in a dependable and timely fashion.[13]

Market Changes that Framed Communications Developments

Manufacturers Hanover serves three basic markets—the American consumer; middle-market companies in fifty states; and a global market of large corporations, government entities, and banks. The institution provides these markets with a range of financial services all over the world. MH is both a truly global and a truly national financial services organization, with $74 billion in assets and operations in forty-four states and all major international financial centers.

The development of Manufacturers Hanover's telecommunications network did not proceed in a vacuum. The institution acts in numerous arenas, all of which had, and have, an effect on telecommunications management. To begin with, there is the world setting. In the early 1970s, financial institutions like Manufacturers Hanover were facing something of a new world order in the increased globalization of business and in the growth—and growing sophistication—of their marketplace. At the same time, deregulation was adding to the number and type of players in that marketplace. The distinctions between kinds of financial institutions were becoming blurred as regulations about interest-bearing accounts were altered or eliminated. Moreover, other kinds of organizations, not part of the banking industry and not subject to the same rules, were beginning to carry out banking-type functions. These two forces—a growing and more demanding marketplace and deregulation—fed each other, and competition grew apace. As someone noted recently, banks have nothing to fear but Sears itself.

MH also works in an industry setting, and the new realities of the 1960s and early 1970s had in many ways revolutionized the culture and image of banks. The widened scope of operations, the demands for new kinds of products, and the market's insistence on new measures of performance all forced a change in how banks did business and in who bankers were.

At the same time, MH possesses a corporate culture, and its staff and activities adhere to that culture. As the range of the institution's competition grew and as the rules of the competition game changed, it became perhaps more important to define the Manufacturers Hanover culture as opposed to the Citibank culture, or the Sears culture, or the American Express culture.

Into this mix of settings came technology, which in the 1960s and early 1970s was opening up new possibilities at a staggering rate. Not everybody understood just how explosive the mixture of market change and technological change would be. Nor could everyone foresee, back in 1970, the effect that people like Steve Jobs and Judge Harold Greene would have on financial institutions, with which they theoretically had no connection. It is safe to say that management at MH understood that technology had the potential to make the institution a stronger force in each of the arenas in which it acted. It is equally safe to say that management in general had little clairvoyance as to all the forms that the strengthening would assume. That technology could provide efficiencies and enhance control in an operation's back office—that it had performance power—was understood. But it was only by stopping to take a hard look at the total situation—a look ordered by a visionary senior management and undertaken by some expert technicians—that MH began to understand the potential of technology for communications and what that would mean. In putting it all together, a vision for the future was created, and development of a global telecommunications network began. The vision articulated, and development demonstrated, that the telecommunications network would be a vital element of corporate competitive strategy.

Pre-Network Operations

Telecommunications at Manufacturers Hanover in 1973 comprised two functions. The first was voice, handled by two persons, one of whom contacted the phone company when new phones were needed or old phones had to be moved, while the other analyzed the phone bills and distributed charges to various users. The second telecommunications function was data. Each of the four data centers at that time had a technical support group, and some six to ten staffers in these groups had the task of connecting terminals to mainframes and ensuring that the telecommunications software was operational. This task was called teleprocessing, even if the terminal was connected by a 50-foot coaxial cable to the mainframe down the hall. Telecommunication costs were growing but were embedded in data processing costs and not a discernible issue.

The data centers in 1973 supported what were essentially transaction processing operations. Manufacturers Hanover, like most major money-center banks at the time, had installed computing equipment to serve as faster, more efficient record-keeping machines that could streamline back office operations and reduce the huge staffs and outsize expenses of those operations. This step had led to more on-site data capture and teleprocessing of the data from remote sites to the data centers.

At the time, Manufacturers Hanover's on-line customer applications totaled one—passbook savings. This application consisted of a rudimentary network into all New York branches. Retail customers could come into the branch, and the teller would update their passbooks on on-line terminals.

This was basically the situation when, in 1974, Jack Evans, who today is not only vice chairman but also chief information officer, ordered a review of all MH data processing technology and operational support. This was a business-driven study. Evans wanted an appraisal of the competitive potential of technology, an assessment of how effectively technology-based activities could serve the line units and their customers.

The assessment established that different businesses had different information needs and required different technological responses. For example, the retail market, with its great volume of small-dollar transactions, was significantly different from the wholesale market, which was lower in volume, higher in dollar value, and far wider in scope. Using this market model as a focus, Manufacturers Hanover created separate, vertically integrated operations centers, with each serving a particular business unit. The idea was that the marketing organization would define the needs, whereas the technical support people would create the technological answer. The idea certainly worked; the marketing organizations drove the technical organizations to develop many new applications that brought Manufacturers Hanover ever closer to the customer. The approach soon resulted in four vertically integrated operations groups—retail, wholesale, securities, and corporate for internal needs. Each group had its own hardware and software, its own development and maintenance units, and its own technical support staff. Data communications responsibilities continued to reside primarily within these groups until about 1976.

Because the market demanded faster response time and greater accuracy in data capture and verification, the mid-1970s saw a flurry of activity in transferring more and more systems from batch to on-line. The efficiencies that could be realized through computer interface started a connectivity trend; paper tape connections to the New York clearinghouse and to international clearing facilities could now be effected better and cheaper by computer.

The direct line from market need through the institution's marketing unit to the appropriate data center spurred the creation of numerous cash-management and funds-transfer applications, reflecting the market's globalization and the growing need for financial services other than credit. At first, the customers of these services connected via telephone or telex to terminals in the Manufacturers Hanover back office. These early connections created numerous timing glitches as well as mistakes in terminal input. The decision was made to put the terminal in the customer's office and let him or her key in the data. The timing of the transaction was now at the customer's convenience or need; any mistakes were also his or hers. More-over, a terminal was far more efficient than a telex, and the aim was to achieve a competitive edge through the efficiency and control afforded by teleprocessing.

Many new cash-management and funds-transfer applications were created in the mid-1970s, and soon hundreds of leased lines and dial-in connections stretched out from Manufacturers Hanover to customers' offices to support them. This rapid growth in communications facilities was largely unplanned and had no overall conceptual design; it was an expedient process to meet market needs through technology. As a result, MH soon found itself with twenty-three specialized networks—host-oriented and, in general, each dealing with a unique application. Costs began to skyrocket. None of these networks could communicate with one another. Often, it was difficult to tell—although cus-tomers could—when they "went down." Moreover, each network lasted approximately eighteen to twenty-four months before it needed to be enhanced, made bigger, or replaced; the staff re-quired to handle the work load was eating away at the cost efficiencies the networks were intended to realize. It was becom-ing clear that if technology was to provide a competitive edge in this brave new world of increased demands from without and within, MH had better look in a new direction. From late 1976 to 1980, that new direction for telecommunications was developed.

Creating the Vision, 1976–1980

Senior management had already seen that voice and data con-stituted one entity and needed to be organized and managed as such. Senior management also recognized that this entity needed to be closely linked to the data processing organization. A new group was thus formed, charged with the responsibility for corporatewide voice and data communications, and it was this centralized group that had the task of analyzing MH's com-munications investments and setting a new direction for com-munications development.

The group was charged with achieving economies of scope and scale while remaining responsive to market needs—that is, be-ing able to introduce new products and serve customers directly through managed data communications to keep a competitive edge. How to achieve both aims, with sufficient control for management and sufficient flexibility for change and growth, was the question facing the new organization.

In 1978, with the strong support of the highest levels of corporate management, a feasibility study was undertaken on a managed corporate private data network. The study represented the first real look at the usage of existing data transmission facilities and at the potential future needs of each business unit of the bank in terms of data communications. The results were startling. In early 1979, the analysis showed that some 56 percent of MH's data communications traffic was computer to computer. The bulk of it was in domestic operations, mostly trust and wholesale. International traffic accounted for only 12 percent of the interactive transactions; the metropolitan branch network accounted for some 31 percent.

Yet at the same time, the study team wrote: "In the very near future, expanding customer needs will require that financial information be available in a comprehensive and unified form. Customers will want direct availability of MHC services, data bases, and funds transfer networks via a standardized computer or terminal link that provides a single universal access to multiple Bank services."[14] When the team began to identify those products and services which would be "communications-dependent," the results were overwhelming. From ATMs to investment services, from foreign exchange to personal loan services, the entire gamut of bank business and internal management was covered. The means of information transfer was about to become the way business was done. The medium and the message were going to be the same thing, and what the study team called "a unified corporate network" was identified as the ideal medium.

Design Decisions

The design team considered four alternatives to providing the corporate network. As a base for comparison, one of those alternatives was to continue to proliferate multiple dedicated networks for each mainframe vendor—IBM, NCR, and DEC. Another option was to buy networking services from a network vendor. A third choice was a facilities-managed network capability provided by a vendor. Yet none of these alternatives fully met the strategic requirements. Either the functionality was insufficient or the scope of performance was inadequate, or the result would have been to lock MH into a single kind of equipment or a single approach to telecommunications. The conclusion of the analysis was that only an internally developed and managed network could satisfy all the objectives and requirements that were of major importance to Manufacturers Hanover.[15]

The overall goal was "single universal access to multiple bank services," a network that would be transparent to the customer yet reliable and run at the least cost. An analysis of how to provide those services from "to make" to "to buy" concluded in a series of policy decisions. First, it was determined to purchase rather than rent high-speed, high-capacity technology whenever and wherever doing so made sense. The buy versus rent principle was aimed at insulating MH from tariff increases while enabling large-scale sharing of resources—a corporate data superhighway rather than a complex of one-lane country roads. A second objective was to make use of proven, commercially available technologies. Third was the decision to rely on in-house capability to the greatest extent possible, thus avoiding dependency on vendors or consultants. Finally, it was noted that the regulatory environment might one day provide the opportunity to offer a broader range of communications-based services and products, and should that happen, Manufacturers Hanover wanted to be ready.

In terms of performance and connectivity, especially given MH's global presence, packet switching seemed the most viable option for a technological base for the MH network.[16] Moreover, both ARPANET and Telenet had used the packet-switching technology, so that its viability had been demonstrated in action. After intense analysis of packet-switching technology and capability within the context of the MH strategic guidelines, the decision was made to create a combined private-public data network that would be designed and managed internally and that would be based on X.25 packet-switching technology. The network would use both the hardware and software of Telenet but would be implemented, managed, and controlled by MH. The change was going to be radical, but the study team believed that the approach to it could not be. One cannot overnight convert a large body of market-driven users into contented sharers of a common network. One cannot get corporate management approval—or funding—for a massive change that will affect both the institution as a whole and the customers that it serves. The recommendation was therefore to proceed by stages—from feasibility to design to a sequence of implementation projects, letting each stage prove itself before moving on to the next. It was further decided to test the design decision in a limited pilot.

The wholesale banking center was chosen as the test site. This area was using an outside time-shared service for cash management services, and there were problems. Updating of files and therefore customer service was taking far longer than it should. A small multinode packet network was installed in the data center to test the X.25 potential as a data communications system for three different services. It proved extremely effective in improving communications service and cutting costs. For cash management alone, it was projected to save more than $1 million per year; the product manager requested that it be implemented for all cash management services, especially in the Far East, where the business pressures for cash management were particularly great. It is to this pilot and the follow-on performance of the cash management services that must be attributed a quantum leap in MH's understanding of what the corporate network could become.

Thus, by the end of 1980, a feasibility study and functional specifications had been produced, proposals had been received from appropriate vendors, and the pilot network had been implemented. In addition, a thorough cost-benefit analysis had been undertaken and a pro forma budget was developed to cover a five-year period of network operation; Exhibit 7-2 shows the GEONET cost projection versus actual costs through year end 1986. All these steps would prove a useful basis for expanding the network over the next few years.

As applications were added to the network, expanding its implementation, an ongoing sales effort was also under way to promote the network to both users and management. Strategy papers were prepared, and a series of meetings were held to explain the network, now called GEONET. The analysis projected that the network would flatten the institutional cost curve for data transmission, provide more flexibility for new product rollout, and strengthen corporate strategic positioning. Cost avoidance, not cost reduction, was emphasized. A projection

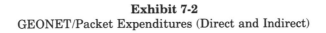

Exhibit 7-2
GEONET/Packet Expenditures (Direct and Indirect)

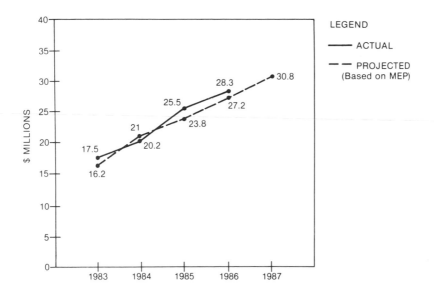

Source: Manufacturers Hanover Corporation.

compared Manufacturers Hanover's telecommunications expense in an unmanaged environment with expense in a managed environment, including the employee base and tariffs. In the unmanaged environment, costs grew due to increased usage and rate increases from 1979 to 1988; in the managed environment, the percentage of increase was projected gradually to diminish to a flat line in 1988. This latter goal was achieved in 1987, demonstrating that the technology could be efficiently and effectively managed, as Exhibit 7-3 shows.

To a business management that saw clearly the growing business need for communications, the "selling" argument was persuasive. The go-ahead for full-scale implementation of GEONET came not as a technical decision from technical staff but as a business decision from businesspeople looking for market opportunities, recognizing information as a product, and concerned about future profits. Similarly, every step of network expansion was a

Exhibit 7-3

MH Telecommunications Expense

Comparison of a Managed versus Unmanaged Environment

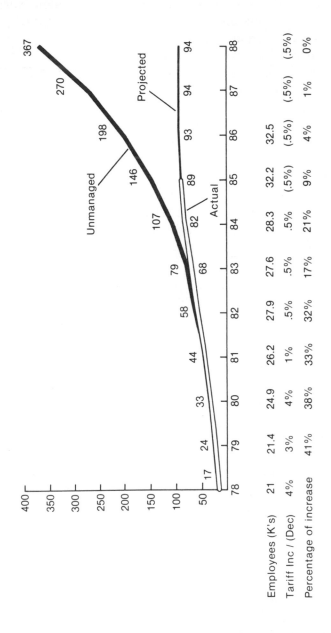

Employees (K's)	21	21.4	24.9	26.2	27.9	27.6	28.3	32.2	32.5	
Tariff Inc / (Dec)	4%	3%	4%	1%	.5%	.5%	.5%	(.5%)	(.5%)	(.5%)	0%
Percentage of increase	41%	38%	33%	32%	17%	21%	9%	4%	1%	0%

Source: Manufacturers Hanover Corporation.

business decision based on opportunity, operating requirements, and the best choice among telecommunications alternatives.

While activities involving data transmission were the most dramatic and visible, parallel activities were under way in the voice arena. Under the leadership of an office systems group, these activities grew incrementally, by replacing existing systems with more cost-effective leased or purchased CMT systems. The activities ranged from increasing the effectiveness of the "plain old telephone service" to implementing sophisticated office automation services, including voice mail. By the mid-1980s voice technology included the installation of many small- and medium-sized PBXs, a private microwave network for voice and data, the purchase of "embedded base" instruments, sophisticated automatic call directors, and other measures, such as least-cost routing and rate stabilization for Centrex services. All these activities had a major impact on the quality of service, functionality, and cost-effectiveness of the voice component.

Implementation: GEONET Today

Implementation—the conversion to the network through the ongoing addition of telecommunications applications—proceeded from 1980 to about 1985. Business pressures demanded broad penetration of first the Far East, then Europe, and then South America. The need to concentrate on the overseas efforts forced a lower priority for the domestic network conversion.

The implementation was not completely hassle-free—not in technical terms, administratively, or politically. Very often, all three kinds of hassles converged in a single issue. Examples of issues that required day-to-day resolution included: software development for unique device handling; security; the interface with worldwide PTTs (the government-run telecommunications agencies of many overseas countries) to gain type approval and permits in each country; the requirement for third-party facilities management in certain sites; worldwide maintenance and parts sparing; hiring and training of engineering, maintenance, and support staff; network management and control; tariff development and chargeback; licensing; packing and shipping of technical products for dispatch overseas; and procuring of local and long-haul circuits composing the network.

The network today is a mixed bag of technologies, vendors,

and combinations of leased and owned equipment. Yet all these things have been by design—and to a design—and MH considers its telecommunications group and its technology state-of-the-art. Peter Keen, in his book *Competing In Time,* commented that Manufacturers Hanover's GEONET was, and is, far better designed and far more cost-effective than Citibank's worldwide communications network.[17]

The GEONET packet-switched system was the communications service that proved that management of telecommunications was possible and economic. It served as the kernel for growth in telecommunications services with respect to voice office support and opened up new opportunities for a variety of emerging data support systems for computer-to-computer transfer of data. A complement to the overseas success of packet switching was an electronic mail project initiated in 1981, when the senior management of the bank, committed to an ongoing, corporationwide productivity program, approved the implementation of such a system organizationwide. The use of electronic mail has grown to include all business sectors, with more than 8,000 subscribers at present. Eighty-five percent of subscribers are professionals, who reap a saving of about thirty minutes a day, or 7 percent of their productive time, from their use of electronic mail.[18]

The bottom line is that GEONET, in terms of both performance and cost, is overall more effective in meeting MH requirements than any of the other alternatives examined. This success may be best exemplified in terms of dollars. With all the mixtures of equipment and vendors, with all the combinations of purchase and lease, with all the redundancies and backups to backups, GEONET is still cost-effective for the institution. In 1986, the total bill of fully loaded costs for the operations of the corporatewide packet-switched network was $28 million. Had equivalent service been purchased from commercial alternatives, that bill would certainly have been higher, based on current tariffs and on cost estimates where comparable services are not available.

The Corporate Telecommunications Market Circa 1987

The deregulation of telecommunications has generated a dynamic market of services that allow a CMT an ever broadening range of options. Supply of transmission services exceeds demand, with several start-up systems affording great bargains and risk. New software switching systems will provide a mutable fare for future innovations with a new level of complexity. With the advent of the 64-kilobit chip, terminal power is leaping ahead. Unfortunately, the change in the "rate of rate" is confounded because of a lack of acceptable standards and an evolving regulatory situation. It is now possible to develop an integrated system for voice and data within a two-year period—*if* a corporation develops internal network standards for data and voice protocols. Nevertheless, the growth of CMTs continues.

The present U.S. telecommunications equipment and service market makes it possible to acquire switching systems and lease transmission service to fit the communications requirements of suppliers and customers at a cost lower than that of current purchased services. This saving is due in part to the rate structure of leased lines and in part to the dramatic drop in the cost of switching and line control systems. As shown in Exhibit 7-4, costs per transmission-mile drop in irregular steps, thereby allowing excess capacity for backup to be purchased while still maintaining lower overall costs. (This swing exceeds the cost of network design and management as well as of service operations.) In those industries in which communications plays such a vital role that firms must manage their own communications services, this opportunity allows an economy of scope for telecommunications activities and increases market effectiveness.

The explosion of new communications products has been a boon to the managers of corporate telecommunications. Accustomed to linking disparate technologies into an overall system tailored to their needs, communications architects can appraise different technologies and connect several manufacturers into an overall system. Air Products, for example, manages a local area network for the company's R&D laboratories that is based on different manufacturers for the transmission, switching, and connection systems.[19] Having detailed information on present and future communications requirements allows the architect to pick the most reliable system for the least cost and, where necessary, to add a chip to tailor it to the company standard. In addition, architects can select the relevant depreciation schedule consistent with the economic life of the product—today three to seven years—and thus appraise the economic consequences.

As competitors introduce such new products as terminals in

Exhibit 7-4
Economic Breakeven Points

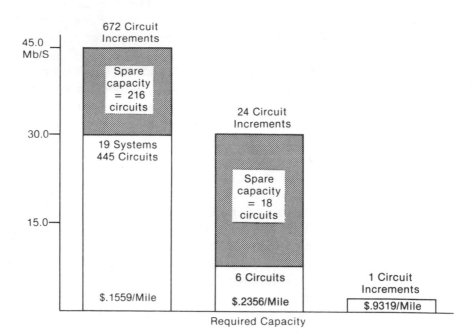

Source: J. McKenney and C. Osborne, "SystemOne: Telecommunications," 9-188-008. Boston: Harvard Business School, 1987.

treasurers' offices, the communications architect is forced to consider both how to meet the competition and how to outflank it with more effective communications for the customer and more efficient communications for the supplier. This situation leads to continuous browsing among innovations and a growing ability to link new technologies to an existing system. Present solid-state designs allow the mixture of different technologies to be judiciously linked in an overall system to take advantage of the power and economics of recent innovations. Moreover, recent business experience has been to prototype systems for test rather than to design and build components to fit the existing system. As one manager noted, "We build quilts by tying things together at the interface rather than by weaving a design."[20]

Finally, maintaining a large internal customer base is difficult as competition grows fiercer among the Bell operating companies (BOCs), AT&T, and other common carriers. Internal customers are not necessarily committed to internal sources. Most large organizations that rely on electronics as a vital aspect of their strategy delegate the management of information systems (I/S) support to the business units. For example, each division of Manufacturers Hanover Retail Banking Trust and Wholesale Banking has its own computer systems capability and can negotiate with corporate telecommunications regarding services. This selection process is not ceremonial; communications costs are a significant influence on a division's ability to provide competitive products. The purchasing power of the divisions thus puts pressure on telecommunications management to deliver cheaper, more tailored, and more reliable service than the available public systems offer.

Corporate Communications Strategies

From a strategic perspective, the importance of communications depends upon the globalization of the corporation, how communications affects the value chain of each business unit, and how the organization has evolved in providing I/S support in the past twenty years. Those companies which were large global users of computer systems—Exxon and Ford, for example—had the first CMT systems. Airlines, financial institutions, and large retailers started managing data communications in the 1970s as their distribution systems became dependent upon electronic links and now have matured to full-line CMTs. Since deregulation, a variety of firms have shifted to active management of their communications as they assume more global dimensions, become more dependent upon communications links, or accumulate or acquire large communications costs.

Deregulation has forced companies to organize in order to manage at least their CPE and to become involved in investment decisions about communications equipment. One initial impact of deregulation was to force many companies to consolidate their communications costs to evaluate how to support customer premise equipment.[21] Most companies had disaggregated communications costs to local geographies, rather than consolidating them corporatewide and thus muddling the total cost of communications. The magnitude of total costs has in several instances galvanized the involvement and action of senior management. At present, a management pursuing the options of alternative service does not have to invent the means; it just has to use the phone, for there exists a broad range of suppliers eager to court the large business user. A large company investigating alternative means of communications services in the fall of 1987 had bids from AT&T, two BOCs, MCI, GTE-Sprint, Boeing, and a railroad, as well as from several service companies seeking to be the prime contractor and to purchase service or equipment for the company.[22] All were forecasting costs significantly lower than the regulated purchased service charges.

All segments of the value chain from R&D to service are being influenced by telecommunications. Air Products has a network for its researchers that allows them to link, through their terminals, to their experimental equipment. In this way, they can set up and control the experiment, move the resulting data to a computer, review the analytical results on the terminal, and decide what to do next—in short, a researcher can manage an entire research activity from his or her desk. Air Products' network has achieved significant improvements in productivity in its research program. McDonnell-Douglas has linked its computer-aided design (CAD) systems to those of its suppliers in order to coordinate the design of new products. General Motors, linked to several of its suppliers, can automatically notify them of schedule shifts or revisions that influence their production. GM's upholstery suppliers in North Carolina have an on-line scheduling system that relies upon GM's daily production schedule; both supplier and head office agree that the system has provided better service at reduced costs. Otis Elevators has a twenty-four-hour-a-day computer communications service for elevator repair.[23] And Westinghouse has an on-line link to all its turbines to provide instant maintenance support, as well as product improvements.[24] In sum, in each aspect of the value chain, from R&D to service, CMT can be shown to be a vital function that enables a range of strategies, from cost control to product differentiation.

Developing a CMT Strategy

As the Manufacturers Hanover illustration demonstrates, the fundamental step is similar to Vail's leadership of AT&T to control service end to end. Management must be committed to a long-range objective of managing communications to provide economical service. This step depends on a belief that the strategic value to the corporation warrants the venture and an understanding of the high fixed costs and volume sensitivity of communications economics.

The early entrants into the CMT business backed into communications management through data processing and had a technological imperative, given their objective of logistical control. In addition, the technological problem was complex, requiring many years of support to be successful. In general, the rewards were impressive; the inventory savings and customer service at Ford and the overall logistical savings at Exxon are examples. (It will be recalled that there was no hitch in our oil supply when the Suez Canal was closed; the CMT worked overtime on that one.) These early innovators now provide evidence of the long-term strategic and economic value of CMTs.

The development and implementation of a communications strategy should be approached cautiously and with competent technical support. Communications management is a new business to most firms, and management must learn to set reasonable objectives, to develop designs that fit their business, and to implement a linked system they can operate well. The resources deployed can range from all purchased to 90 percent owned, although the latter method is probably unique to railroads, as most transmission systems will be leased. A necessary condition is that all telecommunications systems be actively managed to ensure quality of performance.

The objective of a CMT is to deliver quality telecommunications services that meet the needs of the business in a least-cost method. By investing in telecommunications, an organization can improve service over the competition and still maintain cost control. Managing telecommunications can produce more predictable costs that can be controlled over the long term.

The three network components—transmission system, switching system, and local distribution—are all different as to investment and economic life. Local distribution systems are the least affected by recent technologies; here, the most significant cost is the labor required to install the system. Most firms, therefore, as they install the distribution systems, invest in more capacity than needed—for example, optical wave guides before they are usable, as these are depreciated over an extended period and their cost thus becomes negligible.[25] Switching systems at present are rapidly improving in functionality and are going down in price as they are becoming sensitive to the power of microelectronics; they thus will continue to improve in cost performance for the immediate future.[26] This trend is expanding the nature of communications services by offering new services not hitherto supported—combined voice and data or the addition of video, for example—at costs lower than those of existing services.[27] The cost savings and/or productivity improvements continue to make switching investments so attractive that, for some organizations, the constraining function is the ability of the organization to implement and then support the service.[28]

The most critical resource in the development of a reliable, adaptive CMT is a qualified team that is technically competent and managed to support the business objectives. Present CMTs must be managed to continuously adapt to changing technologies and market forces—a new modus operandi for most communications specialists. The role of the technicians is to frame the technical alternatives. Network design is often based upon the technical alternatives, that is, economic scenarios to evaluate and decide on the basic architecture and what services are to be developed or purchased. The latter depend upon how active the firm intends to be in the life cycle of system development—design, construction, operation, and maintenance. Most mature CMTs do all four activities.

Exhibit 7-5 lists how Hewlett-Packard acquires its telecommunications, outlining the company's approach to providing service to four distinct geographical entities: international, national (interLATA), intraLATA, and intraorganizational sites. For each of these areas, three major sets of choices must be

Exhibit 7-5
Hewlett-Packard Sources of Telecommunications Support

1. Purchased public switched voice networks, U.S. and foreign.
2. Leased private lines—data and voice transmission.
3. Owned packet switches for X.25 network based on leased private lines.
4. Purchased X.25 packet service from Telenet in the United States and overseas.
5. Leased T-1 links for large-volume links—data, TV, and concentrated voice.
6. Owned data communication switches.
7. Owned microwave system for voice and data transmission to local sites.
8. Owned high-speed local area network.

considered: transmission capacity, switching systems, and means of distribution to local users. For example, a typical start would be to lease a set of lines within a LATA for intraplant communication; these lines would connect to owned PBXs and terminals linked to a local area net to provide local communication service. The next step would be to lease national lines from a carrier and/or perhaps build microwave circuits between plants. Experience to date has shown that, over time, the geographic base of the system grows to accommodate marginal opportunities to substitute internal service for purchased service. To be able to link disparate systems, signaling standards must be established that allow the interconnectivity for each service.

A New Vision for Telecommunications and for CMTs

Corporate needs are changing, and the changes give rise to the possibility that managing telecommunications is in a time of transition. As effective as most CMTs are, they are not necessarily the final configuration for telecommunications systems; rather, they may be only the beginning.

The forces for change today are not dissimilar from those faced in the 1970s. Market needs, coupled with expanded technological capacity, are the driving forces of change. An example is Manufacturers Hanover's GEONET; it continues to expand as an electronic highway, delivering ever broader and more complex financial products. Being able to get information to a customer earlier than the competition or giving a customer a more useful format for analysis has great bearing on competitiveness.

Emerging technologies are also a force for change. In the future looms the integrated services digital network (ISDN), and the question is whether its promise of more capability, functionality, and flexibility will be realized. Also on the horizon are such technologies as expert systems, imaging systems, and database processing systems. These technologies may so change processing performance that they alter the infrastructure of information technology.

Equally significant are the realities of the environment in which organizations now find themselves. Today, as was the case in the early 1970s, many organizations, especially banks, are in a profit squeeze. They are faced with the challenge of wringing more and more cost-effectiveness out of telecommunications than ever before as they continue to use telecommunications for competitive advantage. They must find better ways of managing this important corporate asset, and that managing, in turn, may have an impact on the structure, scope, and shape of the CMT organization.

One issue that continues to be troubling for the user is the lack of telecommunications standards for network development. The tensions in the international community and between the computer industry and the communications industry, coupled with technological uncertainty, have thwarted most efforts toward the establishment of standards. This situation has forced individual corporations to define their own standards, to which they then adapt purchased and leased equipment. Foreign competitors are beginning to develop operational standards over a broad range. There is a danger that U.S. suppliers, taking a stance of noncooperation for reasons of competitiveness, may thus abdicate standards development to foreign interests. Such abdication results in an uneven playing field, with U.S. competitors stuck in the low end of the field.

In the 1970s, one of the spurs to the development of the CMT was the failure of the then-monopoly carrier to do the job adequately. Today, it is necessary to explore how much the telecommunications vendors have learned since then. Are they smarter now? Are they looking for more opportunities to sell products? There are indications that they are; indeed, any number of vendors now indicate their intention to enter the full-service network maintenance market. Still others are talking about worldwide networking capabilities and/or facilities management. Can the vendors now begin to use their excess capacity and resources to provide a global network cheaper and better than large users can provide by themselves?

Large users are well up the learning curve when it comes to networking; that means the vendors will have to provide a level of service and cost-effectiveness that large users know they can provide. Present-day CMT operators know what problems confront network managers. Few large users will completely abdicate responsibility for their own networks. They will demand a hand in the control and management of their networks. However, they may subcontract for certain vendor services and capabilities where they deem it feasible to do so.

Over the past twenty years, most large businesses have made

a significant investment in learning the telecommunications business. Although most have found it more complicated than they assumed, it has proved not only manageable but in recent times a source of positive cash flow. In the process, most industries have become strategically dependent on tailored systems that can adapt to market shifts and can be orchestrated to enable a product strategy. Several have a customer base that, given present price structures, allows them to make profitable investments in expanding their telecommunications services. As this structure seems in large part dependent upon emerging technology, the large business will continue to be a strong participant in the competitive market for providing communications services.

Notes

1. M. Porter, ed., *Competition in Global Industries* (Boston: Harvard Business School Press, 1986).

2. "Progress Toward International Data Networks," *EDP Analyzer* (Vista, Calif.: Canning Publications, January 1975).

3. A.G. Oettinger, P.J. Berman, and W.H. Read, *High and Low Politics: Information Resources for the 80s* (Cambridge, Mass.: Ballinger, 1977).

4. American Hospital Supply Corp. (A), "The ASAP System," #0-186-005. Boston: Harvard Business School, 1986.

5. P.G.W. Keen and R.G. Mills, *Stages in Managing Telecommunications* (Lexington, Mass.: Nolan & Norton Study, 1982).

6. E. Hall and D. Arthur, *A Methodology for Systems Engineering* (Princeton: Van Nostrand, 1962).

7. H. Borko, "National and International Information Networks in Science and Technology," in *AFIPS Conference Proceedings,* vol. 33 (New York: AFIPS, 1968), part 2, p. 146.

8. "Progress Toward International Data Networks."

9. "Manufacturers Hanover Trust Co.: Office Automation Group," #9-183-049. Boston: Harvard Business School, 1982.

10. J.L. McKenney and M. Vitalie, "Continental Bank," #9-183-044. Boston: Harvard Business School, 1983.

11. "Keeping Abreast of Telecommunications," *EDP Analyzer* (Bethesda, Md.: United Communications Group, March 1986).

12. Manufacturers Hanover Corporation, "High Technology: Banking's Not So Quiet Revolution," in annual report, 1984, p. 34.

13. J.F. McGillicuddy, "Corporate Treasury Management and the New Banking," address before the National Corporate Cash Management Association, New Orleans, November 11, 1985.

14. Manufacturers Hanover Trust, *Corporate Network Feasibility Study,* (New York: MHT, March 1979).

15. J. McKenney and J. Sviokla, "Manufacturers Hanover Corp.: Worldwide Network," #9-185-018. Boston: Harvard Business School, 1985.

16. J. Martin, *Future Directions in Telecommunications* (Englewood Cliffs, N.J.: Prentice-Hall, 1975).

17. P.G.W. Keen, *Competing in Time* (Cambridge, Mass.: Ballinger, 1986).

18. Manufacturers Hanover Trust, "Electronic Mail Recommendation for 1981–84," October 19, 1981.

19. J. McKenney and A. Wardlow, "Air Products and Chemicals, Inc.: Local Area Networks," #9-187-033. Boston: Harvard Business School, 1987.

20. "Air Products Case Notes." Boston: Harvard Business School, 1986.

21. M. Vitalie and R. O'Callahan, "Compass Computing Services," #9-185-105 (rev. 9/85). Boston: Harvard Business School, 1985.

22. J. McKenney and C. Osborne, "SystemOne: Telecommunications," #9-188-008. Boston: Harvard Business School, 1987.

23. F.W. McFarlan and D. Stoddard, "OtisLine," #9-186-304 (rev. 8/86). Boston: Harvard Business School, 1986.

24. R. Jaikumar, "Westinghouse Steam Turbine Generator Diagnostic System," #9-686-006. Boston: Harvard Business School, 1986.

25. A.S. Mookerjee, "Multinational Computer Networks and the Global Enterprise," unpublished paper, Harvard Business School, 1987.

26. "Communication Trends," *Datamation*, June 1974.

27. *Datamation*, August 1976.

28. J. McKenney and C. Osborne, "SystemOne: Telecommunications," #9-188-008. Boston: Harvard Business School, 1987.

8

Bypass of Local Operating Telephone Companies: Opportunities and Policy Issues

Grandon Gill, F. Warren McFarlan, and James P. O'Neill

Overview

This chapter discusses the recent efforts of several small companies to build a market niche by offering communications service in competition to the regional Bell operating companies (BOCs). This service is the logical successor to the long-distance bypass systems developed by companies such as MCI and GTE-Sprint. We believe there is currently a window of opportunity that makes such ventures attractive and that ultimately, as the best locations are exhausted, this window will close. Substantial consolidation, through merger and acquisition, should then occur in the industry.

To support this conclusion, we will review certain important aspects of the BOC operating environment (services offered, economics, and regulatory climate), discuss different meanings of the term *bypass,* and appraise the impacts of a number of emerging technologies (satellite, fiber, and computer).

Next, we will examine the way a bypass company can operate. Since such companies are just entering full operation, actual experience to date is limited. We will discuss potential sources of revenue, advantages (economic and environmental) in competing with the BOCs, and the potential for building a secure market niche. In addition, we will describe the nature of customers and their incentives to participate in the bypass.

We will then raise the question of attractiveness to the financial community, drawing an analogy between bypass and two industries that in the past have faced a similar environment: the airline and cable television industries. Both experienced a period of rapid piecemeal expansion, followed by substantial consolidation; we will demonstrate the potential attractiveness of such an evolution.

Finally, we will examine the potential social impact of such bypass companies. There is evidence that such ventures may improve the cost, quality, and responsiveness of telecommunications service to some users, at the price of putting a strain on the BOCs in their basic mission to make inexpensive, high-quality service universally available to all users.

Operating Environment of a BOC

This section gives background information necessary to understanding the functions and economics of a BOC and defines key terms.

Services

One effect of the breakup of AT&T was the establishment of seven regional telephone companies (the BOCs). These companies were further divided into geographic Local Access and Transit Areas (LATAs) and were permitted to provide complete telecommunications service only within these LATAs. For calls that crossed either LATA or BOC boundaries, the use of a long-distance carrier (AT&T or other common carrier [OCC], such as MCI or Sprint) was mandated. With this new structure, the telecommunications functions performed by a BOC can be summarized as follows:

- Connecting two intraLATA customers (as in the case of a local call);
- Connecting an outbound call to a long-distance carrier's point of presence (POP), determined by the caller's carrier selection and line availability;
- Connecting an inbound call from a POP to the receiving destination; and
- Interconnecting two POPs of the same or different long-distance carriers.

Exhibit 8-1 illustrates these connections and labels each with the entities currently performing them.

Exhibit 8-1
Telephone Connections

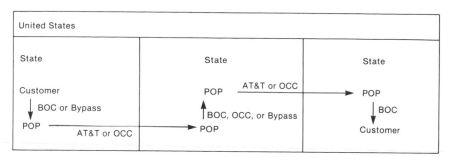

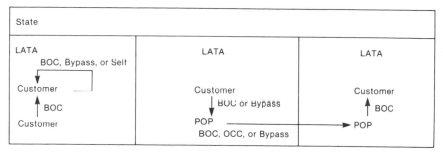

Economics of the BOC

The operation of a BOC is characterized by extremely high fixed costs, with almost no incremental cost per call, also known as "not traffic sensitive" (NTS). This suggests that the most critical aspects of a BOC's short-term economics is its revenue stream. Exhibit 8-2 breaks down total revenue for the seven BOCs by category of service. These elements of revenue reflect the following:

- *Local*
  Flat monthly service charge, typically the major part of the monthly rate paid by businesses and consumers.
  Measured service (message units) in the local area.
- *Toll*
  IntraLATA toll calls completed at both ends by the BOC.
- *Intrastate (but interLATA)*
  Per minute charge on calls made between LATAs, the Common Carrier Line Charge (CCLC).

Exhibit 8-2
BOC Telecommunications Revenues, 1986
(in millions)

	Local	Toll	Interstate	Intrastate
Ameritech	$ 4,528	$ 1,098	$ 1,881	$ 606
Atlantic	4,164	1,422	2,242	369
South	4,881	1,460	2,640	812
NYNEX	5,659	1,254	2,736	248
Pacific[a]	2,957	1,999	2,209	486
Southwest	3,148	893	1,892	906
West	3,259	1,108	2,273	322
Total	$28,596	$9,234	$15,872	$3,749
Percentage	50%	16%	28%	7%

a. Pacific inter-intra breakdown estimated.

- *Interstate*
  Per minute charge on calls originated or terminated by the
  BOC.
  $2.00 per month access charge ($6.00 for business), added to
  the monthly rate.

Regulation

Almost every area of a BOC's business is regulated by the
federal government (through the FCC), state governments, and
the courts (through the Modified Final Judgment associated
with the AT&T breakup). Relative to bypass, the following aspects of regulation are particularly significant:

- *Federal*
  Determines the per minute charge paid to the BOCs by
  AT&T and OCCs for local access on interstate calls. This
  charge is currently set at 3 to 4 cents per minute on originating calls, and about 6 cents per minute on terminating calls.
  Determines the monthly per line charge that customers pay
  for access to long-distance carriers. This charge is currently
  $2.00 per month for residential customers and $6.00 per
  month for businesses. The FCC has requested an increase in
  the residential charge to $3.50 per month by the end of 1988.
  Determines the allowable rate of return on investment subject to federal jurisdiction (crossing state boundaries).

- *State*
Approves the monthly flat rate charge for telephone service (as opposed to long-distance access).
Approves intrastate toll rates.
Approves measured service rates.
Approves intrastate, interLATA CCLC.
Regulates interLATA and intraLATA competition within the state.
Sets allowable return on assets for BOCs operating within the state.
- *Courts*
Determine allowable businesses for the BOCs to pursue.

There is significant variation among states in the degree of regulation exercised outside of rate setting. As of late 1986, thirty-six of thirty-eight multiLATA states permitted long-distance carriers to introduce *facilities* (i.e., physical lines and switching equipment) to compete on an interLATA basis within the state. Normally, this translates to bypassing intrastate calls between area codes. A smaller number of states (fourteen) allow facilities-based competition on intraLATA calls. In these states, carriers can legally bypass any toll call.

There are also widely differing regulations regarding competition and pricing policy on local area services such as Centrex (NTIA, 1986). Such differences influence a large company's ability to connect its facilities with a private network. Their effect on the attractiveness of bypass is negligible, as intraorganizational volume constitutes the bulk of this segment of traffic.

The courts have provided the BOCs with specific guidelines on the businesses they may engage in. In particular, entry into information services, equipment manufacturing, and provision of long-distance services have all been curtailed or eliminated as potential opportunities.

Bypass

This section first describes some of the techniques that fall under the heading of bypass and then focuses on facilities-based bypass of BOCs and the associated technologies and economics.

Background

The accepted definition of the term *bypass* is the acquisition of communications network capability that allows the user to avoid charges he would otherwise incur. Such bypass can occur in two forms: service bypass and facilities bypass.

In *service bypass*, the original form, the user acquires the capacity from the same carrier that he would otherwise use (e.g., leasing a dedicated line from AT&T). This form of bypass can be used to directly reduce the communications cost of a company.

Facilities bypass involves the building of actual transmission systems to handle communications. This form of bypass is economically justified when the costs of installing and operating the system are lower than the carrier charges the system would avoid. Facilities bypass is thus highly dependent upon the local rate structure, and unlike service bypass, it removes all revenue from the carrier being bypassed. It is the impact of this facilities bypass on the BOC that we examine in the pages that follow.

Technologies

A number of technologies are used in performing facilities bypass, the most important of them listed below.

1. *Wire based.* Conventional twisted pair or coaxial cable is used to avoid entering the BOC network. This technology has the disadvantage of being expensive to lay and having low bandwidth (relative to other options), and it typically has no quality advantage over the existing network.

   In the early 1980s, there was the expectation that cable TV companies would attempt to leverage their investment by pursuing wire-based bypass revenues from their customers. Although several companies—especially Manhattan Cable (Gasman, 1986)—have done this, it has not been as prevalent as expected. The companies involved have kept a very low profile. The reasons cited for this lack of participation by cable vendors include the following:

   - Their concerns about the technological complexity of telecommunications;
   - The potential impact of such ventures on negotiations with municipalities awarding cable contracts;

- The differing nature of the majority of cable customers (private, residential) and bypass customers (businesses); and
- The regulations preventing participation that were in place during cable's initial expansion in the 1970s.

2. *Microwave.* Microwave telecommunications are an important part of many OCC networks, and many large individual companies have utilized the technology to reduce internal phone costs. The advantages of microwave are its low cost and high bandwidth, as well as the fact that it eliminates the need to lay cable. Its drawbacks are weather impacts on quality, requirements for clear line of sight, frequency crowding in major metropolitan areas, and, more recently, public perception of possible health effects of excessive exposure to microwave radiation. In fact, however, these latter considerations are not as serious as they seem. Weather has a minimal impact on a well-designed system; frequency ranges allocated to bypass have been expanded and are sufficient in most areas; and there is no medical evidence that the extremely low power levels associated with communications can be injurious to one's health. Microwave is currently the most commonly used form of BOC bypass.

3. *Satellite.* Very small aperture terminal (VSAT) technology and overcapacity in satellite transponder space have combined to make satellite, for many applications, the least expensive bypass technology. The technique involves relaying transmissions between points to satellites rotating around the equator. The major disadvantage of the technology—rendering it unsuitable for most voice applications—is the fixed lag time of approximately half a second that is required for the signal to transit to and from the satellite.

4. *Fiber optic.* Fiber optics is the key technology we examine herein. Like wire-based technology, it is expensive to install, but unlike wire-based systems, it has a potential bandwidth that is so large as to be almost unlimited. This factor makes fiber optics the cheapest technology for most applications, once the traffic volume is large enough. Additionally, the quality of fiber optic transmission is significantly better in comparison with competing technologies,

and it requires far fewer stages of intermediate amplification (repeaters). Fiber optic technology has been widely adopted by AT&T and the OCCs, although it is mainly limited to the high-traffic pipeline between major cities, where the large bandwidth shows the most immediate economic return. Wire and microwave continue to supply the balance of those long-distance networks, with transition to fiber expected to take place gradually over the next twenty years.

Bypass Companies

This section examines the potential opportunities open to a company contemplating a local bypass network. Some of the many factors influencing the attractiveness of a bypass opportunity are also discussed.

Fiber Optic Facilities Bypass

Fiber optic–based facilities bypass involves creating a network that replaces a BOC's "last mile." Doing so allows the bypass company to offer any or all of the following services:

1. *Customer to POP connection.* By connecting the telecommunications user directly to the AT&T or OCC POP, the originating CCLC can be eliminated, allowing substantially reduced rates to be offered to the customer. Alternatively, the service may be sold to the carrier, which can maintain the same rate for its customer while realizing greater margins on the traffic by avoiding the access charge normally paid to the BOC.

2. *POP to POP connection.* The existing OCC networks do not connect to all locations in the United States. This means that calls originating on one carrier must sometimes be completed on another, a situation that may also occur during times of high traffic, when one carrier is at capacity. Both cases require the connection of two POPs owned by different carriers. In addition, the OCCs frequently do not have the facilities to interconnect their own POPs. These connections have traditionally been provided by the BOC, charging an access fee to the OCC. Substantial revenues

are available for a bypass company that provides this service in competition with the BOC.
3. *IntraLATA communications.* Depending on the metropolitan area selected, there are substantial opportunities to
provide dedicated lines for customer use within the LATA.
New York Teleport (NYT), for example, was initially established to provide a concentrating facility for transatlantic
traffic but rapidly found that some of its best sources of
revenue came from connecting together offices of the same
company in downtown Manhattan. Not only could NYT
provide less expensive service, but the company was also
more responsive to customer needs. During a period of
rapid growth in the financial community, this responsiveness was perceived as a major competitive advantage
over NYNEX.
4. *Concentrating long-distance calls.* If the bypass company
has invested in the sophisticated switching equipment necessary, it can concentrate its customers' calls to specific
destinations (such as Washington to New York) and lease
dedicated lines to provide the service itself. By doing so, it
becomes an OCC—by virtue of a service bypass. This revenue is highly dependent on the directionality and predict-
ability of the communications traffic of the customers. If
such patterns exist in an area, the bypass company has a
significant cost advantage over the BOC, which is prohibited by law from such activity.

Influences on the Economic Attractiveness of Bypass

The economics of bypass is highly sensitive to local telecommunications conditions. In conversations with the authors, several industry participants have estimated that there are at least
twenty urban locations in the United States where bypass is
financially attractive at this time. The key assumptions underlying these estimates include the following:

• To motivate customer participation in the bypass, savings of
greater than 10 percent (but probably no more than 20 percent) over BOC rates would have to be offered. Discussions
with the companies and selected customers have indicated
that, in the event that significant quality advantages are

provided by the bypass supplier, the required savings may be as low as 8 percent.

- The companies we interviewed typically use 5 percent of area telecommunications traffic as their targeted market share.

With these assumptions, the attractiveness of a specific location can be determined, in conjunction with the following factors:

1. Total toll volume in the region, with interstate, interLATA, and intraLATA (where allowed) considered separately.
2. Population density, which affects the amount of cable that must be laid to support a given call volume.
3. Market composition, including the following items:

   - Presence of large, multicompany office buildings. These are easiest to connect up, and they frequently contain OCC POPs.
   - Geographic clustering of businesses, as in a downtown business district, which can reduce installation costs.

   Many large companies have already established their own internal bypasses. Their presence must be removed from the estimate of market size.

4. Density of OCC POPs, which both provides an opportunity for interconnection and reduces the amount of cable that must be laid to complete the network.
5. Presence of identifiable high-traffic, long-distance pathways, such as between New York City and Washington, D.C. This presence increases the likelihood that traffic concentration can take place, allowing the bypass company to compete with long-distance carriers.

Advantages of a Bypass Company Competing with a BOC

A combination of regulatory and economic circumstances gives the independent bypass company several advantages in competing with a BOC. For example:

1. *Lower cost base.* Since the bypass company can use state-of-the-art technology, it can build a network with substantially lower cost per unit bandwidth than that of the BOC.

The BOC's ability to react is further slowed through the linkage of its regulated prices to return on assets. This fact has encouraged regulators to mandate extremely long depreciation schedules (up to forty years). Thus, a BOC adopting the new technology might want to make substantial write-offs of existing plant and equipment, an unpalatable item to shareholders, as the equipment written off may be ignored in calculating the rate base. Regulators, on the other hand, may object to the situation wherein the write-off is not made as obsolete and unused equipment may be part of the rate base for years to come.

In addition to having a bandwidth advantage, fiber optics is also nonmetallic. Thus, cable can be laid along power-line runs, at a cost four to five times less than that associated with conventional telephone conduits.

A final cost advantage is the ability to acquire switches, whose prices have followed the plummeting costs of computer technology. The ability of the bypass company to provide sophisticated switching services is vital to competing effectively with the BOC on the basis of quality of service and flexibility.

2. *Provision of long-distance services.* By law, the BOC cannot provide long-distance services within its operating area. Furthermore, in accordance with Judge Greene's Modified Final Judgment (MFJ), the BOCs are not allowed to perform interLATA bypass outside their areas. The bypass company does not operate under such a constraint and can potentially reap from its customers substantial profits that are not available to the BOC.

3. *Customer selection.* The bypass company's most distinct advantage is its ability to construct its customer network solely on economic grounds. This ability allows the bypass company to select only those metropolitan areas which have the most desirable characteristics and to serve only the best customers in the region. The BOC, conversely, is constrained by the necessity of providing inexpensive universal service. It has few economic defenses against the skimming of its most profitable customers; in particular, it cannot lower its rates to special customers without making corresponding reductions in residential rates.

Most Likely Bypass Companies

Given the potential attractiveness of a bypass venture, the question remains as to what types of firms will take advantage of the opportunity. Both the Washington, D.C., and Chicago bypasses are being performed by start-up companies. By regulation, AT&T is prohibited from entering the market. There are, however, other possible entrants:

- *OCCs.* MCI and GTE-Sprint, for example, are logical candidates to enter the bypass business. The major obstacle is their need to direct all available funding to expansion of their long-distance networks.
- *BOCs.* Substantial revenue declines from third-party bypass, although currently prevented by the MFJ, could lead to revisions of the judgment, so that a BOC would be allowed to establish a bypass in the territory of another BOC. These companies would certainly have the requisite experience. The greatest deterrent to such a move, however, might be fear of reprisal from the BOC that is being impinged upon.
- *Other companies.* A variety of other firms, both foreign and domestic, such as cable TV companies, computer manufacturers, and fiber optics manufacturers, could find the bypass opportunity attractive. The major barriers to implementaton appear to lie not with technology but rather with the complexity of site selection, cable placement, and compliance with local ordinances.

No matter who supplies bypass services, the first entrant into a market has a large competitive advantage, for it can skim the best customers and serve them first. Subsequent entrants into the market must either make do with those remnants of the customer base deemed unattractive by the first company or attempt to lure customers away by providing even higher service and lower prices.

Risks to the Bypass Company

The apparent attractiveness of the bypass opportunity should not disguise the fact that certain risks are involved, among them the following:

- If bypass companies were seen to be seriously hurting the BOCs, regulatory environments could change substantially. Business opportunities could be legislated away or tariffs instituted. In such circumstances, compensation or grandfathering would probably be provided to existing bypass companies.
- Costs of network installation are not perfectly predictable. Rights-of-way must be secured, cable laid, and switches acquired. Most bypass start-ups have only limited experience in estimating these costs.
- Some targeted customers will emotionally resist giving up their local phone company, or will demand unreasonable savings.
- BOC response to incursion could include imaginative changes in rate structure that make it harder to compete.

The fact that several start-ups have already been funded suggests that these risks are not inconsistent with the level of opportunity.

Bypass Customers

Crucial to the economic feasibility of a facilities bypass is the availability of a strong customer base. This section first briefly describes the Washington, D.C., and Chicago networks currently under construction and then examines the incentives for a customer to participate in a bypass.

The Washington, D.C., Bypass

Institutional Communications Company (ICC) commenced operation of its network in early 1986, selecting Washington, D.C., because it has the highest concentration of business telecommunications users and carrier POPs in the United States. The network was initially established in downtown Washington, D.C., with connections made to the Pentagon (the largest office building in the world) and plans for expansion into the Baltimore area under way.

Within the next few years, the estimated toll volume originating and terminating in the region is expected to grow to approximately $2 billion. We estimate that the cost of installing the bypass facilities will range from $25 to $50 million, with capital derived from a mix of debt and equity. The amount of cable to be laid in order to complete the network is in the range of sixty to eighty miles.

ICC's strategy centers on gaining revenue from interconnecting POPs and providing intraLATA communications services between large office buildings, particularly federal agencies like the IRS. The company has commenced service and indicates that progress has been satisfactory—although, naturally, costs have been higher than expected and market penetration slower.

The Chicago Bypass

Chicago Fiber Optics Corporation (CFO) is currently constructing a fiber optic network that will service downtown Chicago. According to the city council's minutes, just under twenty miles of cabling will be installed.

The Chicago market is attractive because it serves as a major hub for cross-continental telecommunications traffic. Industry experts estimate that it ranks second only to Washington, D.C., in terms of the number of carrier POPs in place. In addition, a number of communications-intensive activities, such as those of the Chicago Board of Trade (CBT), are present in the region. The cost of installing the network is estimated at between $15 and $25 million.

CFO's strategy is similar to ICC's. Major users, such as CBT, have been targeted initially, to generate substantial early revenues. Additional customers will be added later. There is no indication at present that CFO has plans to enter the long-distance market.

Customer Incentives to Participate in BOC Bypass

The impact of the bypass on a company's cost of telecommunications will be a major determinant of whether a company will participate in a bypass of its BOC. Other factors, however, influence the degree to which costs must be reduced to make bypass attractive. These factors include:

- *Inability to implement internal bypass.* A number of very large corporations can justify building their own internal bypass network. GM, for example, will soon boast the largest private network in the world (Gasman, 1986). Thus, it is medium- to large-sized companies, often housed in large office buildings with other companies, that gain the most from a bypass company. Government agencies also fall into this category, owing to the political sensitivity of state-owned bypass of private sector services.

- *Quality considerations.* The high quality of fiber optics provides significant inducement for customers to bypass, particularly in a world of digitized data communications. The advantage is particularly large for customers with intra-LATA needs. This incentive increases as data grow as a percentage of telecommunications traffic.
- *Carrier response.* Currently, many BOCs face severe backlogs in order fulfillment for new customer digital services. In the extreme cases, delays range from three months to more than a year (admittedly low by international standards). All the bypass companies surveyed cite rapid response (measured in days) to customer requests as one of their key strategic goals. Such response is particularly important for customers whose telecommunications needs are expanding rapidly.

In addition to benefiting the users of telecommunications services, the advantages of the above factors also accrue to the OCCs. GTE-Sprint, for example, advertises its goal to build an all-fiber network. The quality advantages gained from accomplishing this goal are severely reduced if many of the transmissions are subject to switching through BOC wire networks. Similarly, if a bypass company can provide rapid response to OCC switching needs, dramatic savings for the OCC will occur through the reduction of access charge costs in rapidly growing networks.

When MCI and GTE-Sprint initially entered the long-distance market, they estimated that cost reductions of approximately 20 percent would be required to induce customers to select them over AT&T. As public perception of MCI's and Sprint's quality grew, this differential was steadily reduced to its current level of just over 10 percent. As a bypass company demonstrates significantly higher levels of quality and response than those provided by the BOC, the required savings targets decline rapidly. Discounts in some areas are already as low as 8 percent.

Financing the Bypass Industry

A major factor affecting the rate at which bypass occurs is the availability of funding for start-ups. Two aspects significantly impacting this attractiveness are the following:

- *Liquidity:* The expected availability of purchasers for bypass equities in the future.
- *Premium:* The likelihood that bypass equities will command a premium over market values placed on earnings and assets.

To evaluate these factors, it is vital to have a view of how the industry will evolve. The current state of bypass is too embryonic to provide good clues. Thus, assessment must be performed by analogy to similar industries.

A number of characteristics, previously discussed, were used to screen candidates for this analogy. Those characteristics included:

- Having a high fixed-cost, low variable-cost structure;
- Being technologically sensitive;
- Being facilitated by regulatory changes; and
- Having existing entrenched competitors.

Two industries we surveyed shared three of these qualities. Their courses of evolution are surprisingly similar.

The Newly Deregulated Airline Industry

Between 1976 and 1978, major relaxation of regulations governing air carriers occurred. Within three years, a significant number of new airlines were created. These new airlines were similar to bypass companies in that they were created as a result of regulatory change, operated with a high fixed-cost and a low variable-cost structure, and faced entrenched competitors significantly larger than themselves.

The evolution of the industry, freed from regulation, was driven by a dramatic cost differential between existing major airlines and the start-ups and commuter airlines. For example, Meyer and Oster (1984) estimated that differentials in flight crew costs (wages and benefits) equated to a 10 to 15 percent advantage for the new airlines. Total cost advantage (including ground crew and facilities) was as large as 25 to 30 percent. Upon the new airlines' initial entry, higher passenger utilization of flights led to a total cost advantage of as high as 50 percent, although such revenue advantages were quickly removed by the competitive response of the larger carriers.

After a period of intense competition in the early 1980s, however, a wave of consolidation hit the industry. Between 1985 and 1987, numerous mergers and acquisitions took place, among them, People's Express-Frontier; United-Pam Am (routes); TWA-Ozark, Northwest-Republic; Texas Air-People's Express, Eastern, and Continental; American-Air Cal; Delta-Western; and USAir-Piedmont. Few of the new carriers have survived. Airlines like Midway, with a unique geographic franchise, are the exception. The return to the initial investors was critically influenced by the financial strength of the airline at the time of the takeover.

The Cable Television Industry

During the 1970s, there was an explosion in the growth of cable television, an event facilitated by a number of technological advances, especially in microwave and satellite communications, and aided by significant changes in regulation. Like those in the bypass industry, cable operators faced existing competitors (networks and independent stations), had low variable costs, and provided a distinct quality advantage in many areas, in terms of both reception and number of stations offered.

The industry evolved from the development of master antennas in the early 1950s. These systems were designed to provide clear pictures to remote locations in fringe reception areas. Concerned by the possible deleterious impact of this technology on the small UHF stations it was trying to encourage, the FCC brought all cable television operators under its jurisdiction in 1966. This action was a "de facto ban on further cable growth in the top 100 urban markets" (Hollins, 1984). Cable growth stagnated until the ban was removed in 1972, at which time there were 1,570 systems serving 1,575,000 households, about 3 percent of U.S. television households.

After the ban was removed in 1972 and standards were created by the FCC, the industry experienced compound growth in excess of 20 percent per year for the next decade. Toward the end of this explosive growth period, rapid consolidation took place, with the owners of the vast majority of systems receiving an attractive rate of return on their initial investment (site location and market penetration being the major determinants of this return).

Implications for the Bypass Industry

Both the airline and cable TV industries followed a similar path of growth, characterized by these elements:

- Initial entry by a limited number of small companies. Level of success acceptable but not extraordinary.
- Explosive growth in the number of participants, with strong profitability.
- Rapid consolidation of the industry, led by the most successful start-ups, investors seeking diversification, and original industry participants. High premiums frequently paid for less successful competitors.

Such a scenario is very attractive to potential investors. The existence of a consolidation phase facilitates "cashing out," even during periods of lackluster performance by the stock market. It also reduces the risk associated with funding a weaker company, as the expectation that a premium could be obtained is strong.

In addition to the obvious similarity of the analogous industries, there is good reason to believe that such an evolution will occur for bypass. Rapid growth in data communications will lead to increased demand for quality service; regulatory environments are generally becoming looser, at both state and federal levels; and switching technology will continue to decline in price, reducing the minimum network size for which bypass is economical. All these trends facilitate industry growth. Furthermore, there is synergy associated with consolidation of bypass companies, allowing them to better exploit the long-distance opportunity. Currently, the OCCs are utilizing all their available resources to complete construction of their nationwide networks. Once the networks are complete, however, there will be strong incentives for them to purchase bypass operations, especially where such operations are connecting their own POPs. BOC participation may also be legal by this point. The strength of the bypass operation will, of course, be a critical factor in the premium that can be commanded.

Naturally, increased state or federal regulation could dramatically change the scenario. The final section of this chapter will examine the social costs and benefits of bypass, along with possible regulatory stances.

Policy Issues and Conclusions

Both costs and benefits are associated with allowing unregulated BOC bypass ventures to flourish. Below we examine them separately and then summarize our key conclusions.

Social Costs of Bypass

Given the social desirability of universal inexpensive telephone service, any venture that removes revenue from the BOCs without contributing to this goal must be considered carefully. Facilities bypass is a clear example because, unlike service bypass, it removes revenue from the BOC instead of redirecting it from access charges to leased-line fees.

The degree to which BOC revenues are endangered by bypass will be highly location-specific. Approximately 80 percent of all toll traffic is attributable to 20 percent of the business customers of the phone company. In many areas, such as the state of Illinois, a few large customers account for almost a third of all toll revenues. Industry sources estimate that more than half this cost is returned to the BOCs in the form of access charges. Given the selective nature of bypass and the existence of private networks for extremely large users, no estimates of the threat of third-party bypass exceed 20 percent of total inter-LATA and interstate charges. Ten percent is a more reasonable upper bound if only outbound traffic is considered. Exhibit 8-2 shows this level of penetration would equate to approximately 3.5 percent of total revenues in the worst case, which is not inconsistent with Jackson and Rohlfs's (1985, 1986) estimate that each 1 percent of bypass adoption (including WATs) would equate to loss of 0.6 percent of access revenues, equivalent to about 0.3 to 0.4 percent of total BOC revenues.

Roughly 50 percent of BOC revenues is derived from monthly charges. Estimating the average residential connection fee at $15.00, this 3.5 percent of revenues would translate to approximately a $1.00 per month tariff increase for the average customer.

If significant service bypass of the OCCs and AT&T occurred, some impact on long-distance revenues would also be experienced, although mollified in that bypass lines would have to be leased from these same carriers.

Our analysis suggests that the ultimate impact of BOC bypass on consumers would be measurable. In the event that such an outcome were considered undesirable, regulators would have two mechanisms available to make bypass less feasible:

- *Decreasing CCLCs in favor of higher per month fees.* This approach would eliminate most or all of the economic incentives for bypass. Currently the FCC has stated its intention to raise the residential rate to $3.50 per month by the end of 1988. That figure is still well below Jackson and Rohlfs's (1986) estimate of $8.00 per month required to eliminate all incentives for bypass.
- *Increasing regulation of interLATA and intraLATA carriers.* Assessment of the likelihood and timing of this prospect is one of the great imponderables facing bypass companies.

Benefits of Bypass

As previously noted, the customer should see major benefits of bypass in the form of increased carrier responsiveness and improved telecommunications quality. There is a corresponding social benefit: low-cost development of an improved communications infrastructure. The existence of such an infrastructure may allow customer firms to compete more effectively in their individual industries. This perspective suggests that the entry of bypass companies can be viewed as a means of stimulating economic growth outside the telecommunications arena. Given that the U.S. economy is becoming increasingly driven by information and service, this benefit is not trivial, although it is exceedingly difficult to assess.

Bypass permits these benefits to be achieved in a politically expedient way. If, for example, a BOC were to petition to build a fiber optic network de novo, substantial resistance would be incurred because increasing its cost structure would be tantamount to providing justification for a future rate increase. Although these petitions have occurred in some urban centers, the approval process has been a slow one. By allowing outside companies to build the infrastructure, the same rate increase is likely to come about (owing to higher BOC costs), but the direct cause and effect would be less attributable to the regulatory agencies involved. Thus, the bypass permits these services to be introduced more rapidly.

Conclusions

The establishment of facilities bypass companies using new fiber optic and switching techniques poses important and complex issues. Our interviews, particularly with companies either in the bypass business or about to enter it, leads us to conclude that:

- Fiber optics bypass is a feasible and economic opportunity in a number of locations, particularly dense metropolitan areas with high traffic in telecommunications.
- In many states, regulatory and rate environments promote rather than discourage such ventures.
- The major customers of bypass include both medium- to large-sized firms and long-distance carriers (OCCs).

Our analogy to the airline and cable television industries suggests that an investment in new bypass technology companies may be quite attractive. Unless there is government intervention, we expect to see a significant rise in the number of bypass start-ups, followed by a period of consolidation. Most bypass activities will not ultimately emerge as independent entities.

With respect to social aspects of bypass, we conclude that:

1. The benefits of bypass to customers include quality and responsiveness of service as well as cost savings.
2. Extensive bypass activities will endanger some portion of the BOC rate base, without a corresponding reduction in its cost base. This creates social costs:

   - A larger portion of the telecommunications cost burden will fall on consumers and small businesses. The effect will be measurable.
   - The BOCs' attempting to compete with bypass companies will cause further deterioration in service to customers that are not bypass candidates.
   - The cost of the "lifeline" ideal of low-cost phone service

for all will grow as the "cream" of the BOCs' business is skimmed.

3. The social benefits likely to accrue from bypass activities include:

   - The building of a technologically advanced, high-capacity infrastructure in urban areas.
   - Increasing levels of telecommunications quality available to business customers, enhancing the competitiveness of the customers.
   - An upgrading of the transmission quality of OCC communications hubbed through the bypass.
   - An increasing BOC incentive to be responsive to customer needs.

By altering rate structures and competitive regulations, state and federal agencies can completely determine the magnitude of the bypass opportunity. We believe that such power should be used only to the extent that it:

- Encourages those types of ventures which will build a communications infrastructure that expands the range of options available to customers competing in an information-intensive world; and
- Reduces the desirability of those efforts which are mainly offspring of the rate environment, whose benefits are solely the result of reducing OCC and business costs at the expense of a BOC's other customers.

References

Gasman, Lawrence. "The Bypass Connection." *High Technology* (May 1986).

Hollins, Timothy. *Beyond Broadcasting: Into the Cable Age.* London: BFI Books, 1984.

Jackson, C.L., and J.H. Rohlfs. *Improving the Economic Efficiency of NTS Cost Recovery.* Washington, D.C.: Shooshan and Jackson, 1986.

———. *Access Charging and Bypass Adoption.* Washington, D.C.: Shooshan and Jackson, 1985.

Meyer, J.R., and C.V. Oster. *Deregulation and the New Airline Entrepreneurs.* Cambridge, Mass.: MIT Press, 1984.

National Telecommunications and Information Administration (NTIA), Office of Policy Analysis and Development. *Telephone Regulation and Competition: A Survey of the States.* Washington, D.C.: U.S. Department of Commerce, October 1986.

Part Four
Electronic Information Services

9
Enhanced Communications Services: An Analysis of AT&T's Competitive Position

Joseph Baylock, Stephen P. Bradley, and Eric K. Clemons

Introduction

As a society, in all of our daily activities we are becoming increasingly reliant on enhanced communications services. These enhanced services merge communications and information processing in novel and powerful ways. Services that generated a good deal of wonderment five years ago have become not just commonplace but expected. We have all felt a tinge of frustration while waiting for a salesclerk to consult a printed catalog of stolen credit card numbers, or perhaps full-fledged anger when an automated teller machine refuses to dispense much-needed cash because the telecommunications link has failed.

In addition to these and other banking and retail applications, enhanced services are making inroads in manufacturing, travel, insurance, distribution, transportation, health care, and education, and in the consumer and government markets as well. The list is by no means exhaustive; the notion is that wherever information needs to be exchanged, people and organizations are considering means to do it faster and more accurately, and at a profit.

Their tools can range from a generic PC to custom point-of-sale terminals, from professional workstations to distant mainframes, from batch processing to time-sharing services, from a few keyed entries to multiple database references. The transmission medium may extend throughout the office or the world, it may be private or available to all, it may travel underground or through space, and perhaps most important, it may be owned by the user or any one of a number of providers.

Enhanced services serve many markets in many ways; they include essentially any data communications service that goes beyond simply reformatting and transmitting signals. Many firms view information services as attractive markets based on their size, their fragmentation, their perceived ease of entry, and the prominent past sales and earnings growth figures for certain participants. Various estimates place the worldwide market size at $5 to $10 billion by 1991. The European value-added services market alone may account for $5 billion of this.[1]

The onset of reduced regulatory restraints has prompted AT&T and the regional Bell operating companies (BOCs) to consider how they might enter the enhanced services markets. Historically, carriers have been prohibited from providing the "content" of communications, but *Computer Inquiry III* (CI-III) has defined provisions for relaxing this restriction. Details of CI-III are discussed below. Other new players include value-added common carriers such as GEISCO and US Telenet, which are beginning to forward-integrate toward the applications and databases that have traditionally been carried on their networks. For instance, in June 1987, GEISCO introduced EDI*T System, a software translation package for data generated by clients' applications.

AT&T is newly counted among the players in this industry, and it is, of course, the largest. Since AT&T's founding, R&D, telephone equipment manufacturing, and basic telephone service have been universally acknowledged as among its corporate strengths. After divesting the local telephone operating companies in 1984, AT&T was allowed to enter the general purpose computer market. In its 1984 annual report, AT&T expressed the belief that "the components and electronic systems we produce 'in-house' give our other lines of business significant advantages in the markets they serve." The report went on to say, "In the growing market for business automation systems, our unique strength comes from our long experience in developing and applying both telecommunications and computer technologies."[2] This was a clear indication that AT&T expects to continue to enjoy economies of scope as well as scale in its postdivestiture operations. An electronic messaging system, a videotex information system, and a real estate information service offering were enhanced communications services described in the 1984 report.

At issue is whether AT&T's traditional strength in basic communications services truly carries with it any "unique strengths" or confers any "significant advantages." Financial results show a 6.7 percent decline in revenue for the first quarter of 1987 and a decline of 1.4 percent for the year.[3] AT&T initially predicted profitability from computers in 1988 but now has revised this forecast to late 1989.[4] These results indicate that although three years have passed since shareholders, analysts, and competitors first speculated on the tangibility of these "unique strengths," their significance and benefits remain open to question.

Enhanced Communications Services

Some further definition of enhanced services will aid in the evaluation of AT&T's capabilities. At the most basic level are services that offer alternative forms of message delivery, forms considered enhanced because of the role of computers in the storage or transmission of messages. Included in this category are services such as basic protocol conversion, electronic mail, voice mail, and electronic data interchange (EDI).

Other aspects of enhanced communications services entail a more comprehensive merging of communications and computing technologies. These services allow vendors opportunities for more provision or modification of content, rather than merely message transmission, albeit transmission in a more complex digital form. Accordingly, these services allow vendors more opportunity to increase value and thus to create new sources of revenue. These applications range from providing public databases, through offering on-line transaction services, to providing full interorganizational multiparty information networks.

An alternative classification of the potential market for enhanced communications services is by the nature of the target industry rather than by the technology of the service provided. Certain industries have become more active users of information systems and enhanced services owing to the nature of their businesses. One measure of the diffusion of information technology is computing intensity, as measured by information systems spending (within and apart from management information systems [MIS] organizations) as a percentage of revenue. The Diebold Group has compiled this view of computing intensity:

Industry	Intensity
Banking	4.9%
Retail/sales	3.2
Insurance	2.4
Health care/drugs	2.2
Metal products	1.9
Consumer products	1.9
Electronics	1.6
Public utilities	1.6
Chemicals	1.5[5]

Interorganizational Information Systems

Several companies within these computating-intensive industries have had spectacular successes, in part because they were able to implement interorganizational information systems— systems that ease the flow of information to or from suppliers or customers into the organization. Often these systems have been implemented in conjunction with a third party, a trend that we expect to continue.

American Hospital Supply Corporation shook the health care industry in 1974 when it placed order entry terminals in hospitals. McKesson followed suit in the retail pharmaceutical industry by automating the ordering, pricing, inventory, and shipping processes for its pharmacy customers.

Much of the banking industry's recent computing intensity has been in the form of automated teller machines (ATMs). The 1,425-member Cirrus network links more than 6,500 ATMs and generates 200 million transactions for deposits, withdrawals, and balance inquiries from checking, savings, and credit accounts.[6]

American Airlines' Sabre system and United Airlines' Apollo reservation system have helped increase these carriers' market shares; estimates range as high as 20 percent.[7] The systems became such powerful competitive weapons that accusations of restraint of trade, hearings, and court cases caused them to undergo significant modification to reduce their impact on travel agents' selection of airlines.[8]

Organizations that have not yet converted to an electronic environment and those looking to improve upon the systems already in place continue to fuel the growth of the enhanced communications services markets. The Veterans Administration is currently preparing to put a $200 million contract out to bid.[9] IBM, US Telenet, McDonnell-Douglas Network Systems, and BBN Communications are expected to vie for the right to provide a packet-switching network connecting three hundred VA hospitals, regional offices, and pharmacies. BBN Communications also notes that the rapid pace of mergers and acquisitions in the financial arena, coupled with the proliferation of new financial products, is providing many follow-on opportunities in financial services.

Key Enabling Technologies

We see several new, emerging technologies as critical to competing in the business communications marketplace. These technologies will be essential to any player wishing to provide enhanced communications services for major business customers. We now review the most important enabling technologies.

Digital Transmission

Each of the examples described above makes extensive use of transmission media in connecting origination and destination, that is, customer and supplier. In the past, analog transmission media were the only means available for connecting many locations. Analog channels limit the user to speeds of 4,800 bits per second or less, and quality on the public switched network can vary widely over time and by points connected. Advances in the price/performance of digital circuitry, fiber optic technology, and satellite communications, together with the reduction in maintenance expenses associated with higher capacity channels, have driven the introduction of digital transmission into carriers' networks. These higher speed, higher quality channels, available in a variety of bandwidths, have in turn become a key enabling technology for the economic provision of enhanced services.

Digital Storage

The quintessential example of the declining price/performance curve has been semiconductor memory devices. As capacity and access speeds have grown, worldwide price competition and manufacturing process improvements have led to dramatic price declines. Incorporation of the newer, cheaper memory in computing systems has improved performance across the full range of general purpose and specialized computers. One of the major beneficiaries of semiconductor memory innovations has been the telecommunications industry. Memory-intensive stored-program control switches now form the core of the nation's networks, allowing for improved performance as well as greater flexibility in implementing regulatory changes required in the areas of equal access to long-distance carriers and enhanced services.

Integrated Services Digital Network (ISDN)

AT&T describes ISDN as "a graceful evolution of today's telecommunications network toward a powerful, unified network fabric featuring universal ports, dynamic allocation of bandwidth and adaptive, logically provided services."[10]

ISDN is composed of CCITT (Consultative Committee for International Telephone and Telegraph) standards for:

Access rates
Channel types
Interfaces
Protocols
Signaling

The basic access rate is 144 kbps, configured as two 64 kbps "clear" B channels and a 16 kbps D channel. B channels are considered to be "clear" because they are unencumbered with signaling in the channel. The 64 kbps rate coincides with current standards for digitized voice, but the B channel will also be available for voice at lower bit rates, circuit-switched data at any speed to 64 kbps, and packet-switched data as determined by the signaling information on the D channel.

In addition to signaling, the D channel may be used for remote meter and alarm reading, energy management, and user data up to 16 kbps. The 2B + D16 channel is referred to as a digital subscriber line. Eventually, this 2B + D16 is expected to become the main telephone company offering and the new digital central office standard, just as individual analog touch-tone lines are today's standard.

For users needing more than 144 kbps of bandwidth and for intranetwork node trunking, the CCITT has defined a primary access rate as follows:

23B + D64 channel (in North America)
30B + D64 channel (internationally)

These standards are referred to as an extended digital subscriber line.[11] North American and international standards for extended digital subscriber lines differ because, in the absence of standards for earlier T-1 lines, high-speed communications links have different capacities in the two environments. Just what the implications of differing North American and international standards will be is at present unclear.

Today ISDN is largely driven by technology. Suppliers have cast it as the "service of the future," recognizing that the CCITT originally expected its full development to take twenty years.[12] On the supply side, intensifying competition among local exchange carriers, interexchange carriers, and equipment suppliers for large user accounts is a primary driver of ISDN. It is thus at least plausible that ISDN will be available as a commodity service from numerous vendors, and that ultimately ISDN capability will be widely available and will not be a source of competitive advantage to any provider of interorganizational information systems.

On the demand side, users are aware of ISDN and its benefits. They have a strong interest in it, but they have an overriding responsibility to solve today's business problems. Today's network problems are poor quality and reliability of leased facilities, long lead times for installation and changes, and inadequate network management and diagnostic capabilities, particularly in data networks.[13] These problems are exacerbated by the growth in complexity of the multivendor, multifunction private networks and by the increasing number of users implementing such networks.

One possible scenario views the large and growing number of T-1 networks as a near-term target population ready to upgrade to full ISDN capability when it becomes available; users of these networks may represent a waiting set of innovators ready to adopt ISDN facilities from common carriers. However, the most serious threat to ISDN's success and eventual universal adoption is these users' reluctance to wait for ISDN services. Users with pressing business problems may be unwilling or unable to wait for ISDN. They are pressuring service providers, equipment suppliers, software vendors, and international development teams for solutions over a much shorter interval than the fifteen to twenty years that full ISDN deployment is expected to take.[14] An alternative to widespread ISDN adoption may be further installation of private T-1 networks, with increased support of large-scale computer networks. Clearly, a large and entrenched community of private T-1 users represents a threat to ISDN's role as a critical enabling technology.

An additional limitation of public ISDN offerings may result from the breakup of AT&T and the divestiture of the Bell operating companies (BOCs). To be most valuable, ISDN services should be end to end, yet BOCs and local exchanges may not invest in the ISDN connections needed to complete "local delivery" of digital service. Several of the BOCs have stated that Signaling System No. 7 (SS7) is the last technology they will deploy "on faith."[15] Large users also realize that partial ISDN deployment will not solve their network problems and are particularly concerned about locations outside of major population centers.

CCITT Signaling System No. 7

Common Channel Interoffice Signaling (CCIS) is a relatively new technique for transmitting all the signaling information for a group of interoffice trunks on a dedicated high-speed data network. Although CCIS was only introduced to the AT&T network in 1980, the common channel interoffice signaling format is already being changed to conform to the CCITT's SS7. SS7 offers the advantages of digital transmission at 56 or 64 kbps, a longer and more flexible message format, CCITT approval, and inherent capability with the ISDN access rates. The high- and low-speed D channels form the building blocks for SS7.

The SS7 data network is an overlay network that speeds network management and control functions. It is an important step toward ISDN, a factor that accounts for its implementation by AT&T, MCI, US Sprint, the BOCs, and several of the major independent telephone operating companies. Network efficiency alone probably would not cost-justify SS7 deployment, but all carriers view SS7 as a platform from which a wide array of information services may be launched at low incremental cost. This is an important consideration, because digital switching and transmission, SS7, and ISDN development are all capital-intensive investments. SS7 and ISDN development costs offer significant scale economies; that is, the development costs are largely independent of the size of the network because of the use of software in centralized nodes, which can readily accommodate capacity growth.

The signaling network's flexible message format will support data transfer for custom service signaling, calling card verification, toll-free services, and other services. Since the receiving station receives the telephone number of the station originating the call, numerous potentially valuable custom services are possible. Computer security is enhanced, since dial-in users can be authenticated and it can be verified that the call originated from an authorized location. Marketing and sales support organizations will be able to do market segmentation by geographic location and route incoming inquiries to the appropriate sales group. Furthermore, custom service signaling has been instrumental in immediately and dramatically reducing the number of obscene and crank calls in Canadian cities in which it has been implemented.[16] These are just a few of the new revenue streams that carriers anticipate as enhanced services grow. These new revenue streams should provide sufficient impetus for the long-distance carriers that remain after industry consolidation to offer a full spectrum of such services.

We have described several of the key technological tools that will be needed to provide enhanced services and have shown that economies of both scale and scope will apply toward their deployment and operation. In considering AT&T's position, one cannot doubt the company's technical competence in successfully deploying these technologies. With capital budgets of $2.5 billion in 1987, 1988, and 1989 for expansion of the digital network, AT&T should be quite capable of taking advantage of the economies of scale and scope.[17] Thus, the implication to this point is that AT&T may be well positioned to compete in the enhanced services markets.

The Concept of Strategic Necessity

Recently much attention has been focused on the strategic importance of information systems. However, as we have noted earlier, a system can be strategic for different reasons:[18] it can be a source of *competitive advantage*, allowing a firm to operate more successfully than its competitors, or it can be a *strategic necessity*, necessary for the business but widely available in the industry. A competitive advantage allows the firm to earn above-average profits, or economic rents. Strategic necessities are even more compelling: while getting them right simply preserves parity with competitors, getting them wrong may force the company out of business.

We believe that most strategic uses of information systems will be seen ultimately as strategic necessities. Certainly externally focused systems—those directed at altering relationships with customers or suppliers—rapidly become visible and widely known. Often astute competitors will attempt to duplicate them quickly. In the absence of barriers to entry or customer switching costs, first-mover effects are rapidly dissipated. And all too frequently, competitors can duplicate a retail services innovation faster than consumers adapt it, allowing competitors the opportunity to respond before the market is successfully preempted. In all but a few widely cited instances, such systems have become strategic necessities rather than sources of competitive advantage. And ultimately, as the necessity nature of the product is understood and as more competitors begin to offer it, these systems will become commodity products.

As these applications mature and competitors become frustrated with investment that does not yield sustainable advantage, these commodity services begin to be provided by consortia, rather than being duplicated by each competitor. In retail financial services, ATM networks are an example of this maturation process. These networks are often owned and operated by a consortium—NYCE in New York, or by a single bank like

MAC in Philadelphia. Retail point-of-sale networks for credit checking illustrate another aspect of the process, whereby the consortia turn to a third-party vendor to avoid becoming dependent on any one competitor for vital services.

In the extreme, the player providing the service may abuse the power resulting from the information system, as has been alleged with the airline reservation systems.[19] Even when no abuse of power is claimed, if the information system to support a necessary service were provided by a player in the industry, a competitor would profit with each transaction processed, as is the case with ATM networks operated by an individual bank. Either of these factors motivates the purchase of strategic interorganizational information systems provided by outside third-party vendors rather than by a competitor.

The combined phenomena of strategic necessity and third-party providers creates an environment in which AT&T may very well be able to leverage its technical competence and in-place network. But certainly AT&T is not the only one tempted by the information services business. Ordernet, a ten-year-old provider of Electronic Data Interchange (EDI) systems and a subsidiary of Sterling Software, says that, "a slow year for us is growing by 60 percent. A fast one is growing by 80 percent."[20] Such growth does not go unnoticed and, as noted above, the industry's fragmentation, perceptions of low-entry barriers, and attractive financial performance by some current players, coupled with the large size of the interorganizational information systems (IOS) market, will likely attract many third-party systems developers. They will be joined by companies with viable IOSs looking for product extensions. Foremost among these are the airlines, financial institutions, and owners of vertical market databases. Thus it is appropriate to examine a priori the key strengths a third party must have for success.

Key Strengths Needed by a Provider of Third-Party Interorganizational Information Systems

Role of Technology

One method of identifying the key strengths required to be a third-party provider of IOS is first to examine the particular abilities that will not be critical to success. We believe that having public carrier technical expertise in-house will not be pivotal, because the technology will be widely available in highly competitive markets.

Heated competition will arise from excess digital transmission capacity and ready availability of SS7 and ISDN services or their equivalents from several suppliers. Digital transmission capacity is being installed at the national, regional, and local levels. Nationally, in addition to AT&T's investment MCI and US Sprint are placing extensive fiber optic networks into service. US Sprint plans to have northern, central, and southern transcontinental routes in service by the end of 1988.[21] Further expansion can occur at low incremental cost by adding electronics to unused fiber already in place and by increasing transmission rates. Regionally, Lightnet (in the East), NorLight (midwestern power utility consortium), and the National Telecommunications Network of regional carriers are all operational. Within Local Access Transport Areas, the BOCs and major independents have all been active, as have consortia of smaller independents, such as the Wisconsin network.

From both a technical and an economic viewpoint, a third-party IOS developer will probably be better positioned if it uses capacity from several suppliers. Multisourcing will be advantageous because, as AT&T has stated, "it is uneconomical [for a carrier] to provide separate substitute digital communications routes in case of a disaster or cable cut."[22] Yet some important customers will pay dearly for insurance against a five-hour outage. Thus, multisourcing of bandwidth should yield a more robust, failsafe operating environment.

Finally, very small aperture terminal (VSAT) satellite networks are also experiencing significant growth, adding to the nation's digital capacity. J.C. Penney, Wal-Mart, K mart, Ford Motor, Chrysler, Merrill Lynch, and Citicorp are just a few of the companies that are implementing VSAT networks. Once businesses have made capital expenditures for private VSAT networks, they are unlikely to consider purchasing enhanced communications services from a provider who bundles these services with transport capacity. Furthermore, an enhanced service designed to be bundled with a single vendor's transport facility may not be divisible from the transport facility, thus limiting market acceptance.

It appears that no special advantage will accrue to AT&T in the interorganizational information systems market by virtue of its vertical integration in telecommunications services. Several competitors will offer essential communications services. Providers of IOSs will probably obtain telecommunications services from several sources. And, given the nature of commodity markets with chronic oversupply, purchasers of these communications services may actually enjoy real benefits not enjoyed by providers. Moreover, AT&T could not deny access to its network services to any firm that wished to compete in the market for IOS. Computer Inquiry III requires that AT&T-Communications provide competitors with the same quality of service it provides its own in-house IOS vendors.[23] Instead, it is conceivable that to meet the reliability standards of multisourcing competitors, AT&T might have to divert some portion of an IOS' traffic to competitors.

In brief, we believe that:

- An efficient market will exist for communications services, from numerous vendors; this market will include new digital offerings;
- At least a portion of the potential market will choose to remain with private networks, limiting the impact of expertise in the new communications technologies; and
- Overcapacity will squeeze prices and margins.

When an efficient market exists with serious overcapacity, there is no economic benefit to be gained from vertical integration into that market. This point leads to the conclusion that ownership of technology will be irrelevant as a source of competitive advantage in providing IOS.

Key Success Factors Likely To Be Critical

Returning to the original issue, what are the key strengths needed to be a successful provider of IOSs that are in short supply? We have identified four key areas that we believe will prove crucial:

Systems development and integration
Marketing
Industry-specific expertise
International expertise

Systems development and integration. Critical skills are often in short supply in rapidly growing markets, and the set of systems skills is no exception. The short supply leads to high mobility—the IEEE reports that nearly 14 percent of all software engineers changed jobs in 1986.[24] According to a company officer at American Management Systems, a systems development house experiences a 25 percent annual turnover.[25] AT&T has experienced employee retention problems on past systems development projects. A study of the ill-fated Net 1000 value-added packet-switching project showed that the average tenure of persons developing Net 1000 was less than one year. Howard Frank, a Washington D.C., data communications consultant, notes that, "you had many people working there who were fresh out of engineering school."[26]

High turnover at the entry level exacerbates middle-management retention problems by diffusing the focus on ongoing systems development efforts. Reducing the size and quality of the promotion candidate pool is a second-order problem, which must also be dealt with. This same candidate pool is often the source of talent to meet human resource needs in the next key area.

Marketing. It is helpful to separate marketing issues into those of marketing strategy and those of market planning and implementation. Marketing strategy involves defining corporate and strategic business unit (SBU) objectives, defining and segmenting markets, and selecting a coherent set of strategies for the market segments. Market planning and implementation are concerned with analyzing customers' expected responses to products and using the information for better product planning. Implementation issues range from budgeting, to forecasting, to sales force allocation, and management, and training. Effective market planning calls for finding the relationships among different levels, types, and allocations of resources and the corresponding impact on sales and profits.[27]

We have implied throughout our discussion that the enhanced communications services market is not one market but many markets. And, in fact, each market is composed of a variety of segments. No third-party provider can reasonably expect to compete effectively in all markets because of the wide variations in customers' needs. The successful provider will have the ability to define and segment markets correctly and to match segment needs with its own business strengths. One can expect clusters of segment needs, such as "high reliability and security with low price sensitivity" or "multiple database access across countries with high price sensitivity" to appear. Market strategists will be able to identify and track these clusters of segment needs and synthesize strategies to be passed on to market planners. The analytical skill required for a market strategist put this talent in short supply in most industries—IOS is no exception.

Industry-specific expertise. Packet switching, digital transmission, and ISDN will be commodity services, with little impact on a firm's ability to offer information services. Rather, to effectively offer interorganizational information systems, a provider will need to know intimately the requirements, procedures, and strategies of the organizations it is attempting to support. It will in fact be necessary to appreciate the strategies and needs of individual business units almost as well as their own organizations do, if a vendor hopes to sell them strategic information services.

Connectivity standards, although slow to evolve, will act to lower switching costs, causing more of a focus on the value that is added by the IOS to the customers' specific businesses. And given the visibility and strategic business impact of IOS, the number and type of key decision makers in IOS procurement are increasing.

Current recommended MIS practice places some MIS personnel in the user departments to improve system support and communications with end users, while their centralized colleagues act as distributors, coordinating the acquisition of outside services.[28] It is difficult enough to develop effective in-house information systems without adequate contact with users and without attention to the specific needs of the firm. Developing systems as a third party, with neither industry-specific expertise nor extensive contact with the ultimate users, is far more difficult.

Moreover, marketing and sales become more complex, requiring more contacts and a greater range of skills. As financial, operations, marketing, and even general managers include themselves in the IOS procurement process, they want the benefits to be expressed in their terms. This request is not at all unreasonable, but it does put pressure on the market planner to tailor the product to both the industry and the function, and on sales personnel to know the industry and its participants. Again we are faced with a set of skills that seems in perennial short supply. And finding the necessary people will be particularly difficult for target industries that differ significantly from the information systems industry, for example, consumer goods or health care.

The marketing issues also include implementation, much of which involves sales force management and after-sale support. In an environment where needs, products, and standards are continually evolving, sales force management takes on more and more aspects of relationship management. The customer relationship must be maintained by providing a steady stream of new-product information, training on products recently purchased, and service for the installed base.

Sales force reallocations, organizational restructuring, and sales force turnover are extremely damaging to customer relationships. AT&T has experienced problems in each of these areas. Perhaps most distressing to customers is the turnover in sales force. Since 1982, Abbott Northwestern Hospital in Minneapolis has gone through seven AT&T account executives.[29] Other major customers note a serious deterioration in service in the past two years as the AT&T organization has suffered from continual disarray.[30]

International expertise. Markets are becoming increasingly global. Aside from improved communication, the benefits of international organizational experience can be seen in heightened sensitivities to cultural differences, greater appreciation of risk (political, foreign exchange, and so forth), improved regulatory and market intelligence, and a better understanding of the possible uses of counter-trade. A company with a track record in other countries may be more open to joint ventures and foreign sources of capital while being less susceptible to nationalization efforts. In some lines of business, international expertise and presence may be of little importance. But there can be no doubt that in serving many of the computing-intensive areas an international perspective will be a key strength that providers of interorganizational information systems must have.

Many companies, like Reuters, now believe that there are significant opportunities for information systems to support international securities trading. Clearly, success in this marketplace will be determined more by financial expertise, systems development strengths, and access to information than by access to communications technology.

Competitors Facing AT&T in the Development of Third-Party IOS

We have identified several factors that have attracted or will attract new competitors as the enhanced services market develops. We shall examine several of these key competitors facing AT&T, focusing on their business strengths as measured against the key success factors.

Systems Developers and Integrators

Several systems vendors are already competing for the market into which AT&T hopes to move. Many of them have enormous resources, extensive industry contacts, experienced sales forces, and considerable experience with telecommunications networks. We highlight here only a few of the organizations confronting AT&T as it contemplates moving into this market.

EDS Communications Corp., Bethesda, Maryland. EDS, a subsidiary of General Motors, currently has more than 2,000 systems professionals on staff. It was founded by ex-IBM supersalesman Ross Perot, who instilled his and IBM's marketing orientation at its inception. EDS offers contract systems development and integration work—most notably for AT&T in structuring competitive bids—as well as information services via its in-house network. EDS*Net serves twenty-five nations, with voice and packet-switching provided by 270 switches and 500 packet assemblers and disassemblers (PADs) connecting three-quarters of a million devices for 4,400 accounts. Although the primary backbone is AT&T, US Sprint has made significant inroads with digital fiber, and the international links are expected to be provided by sister GM subsidiary Hughes Communications Corp. Hughes has an international focus, a strong track record of successful systems development, and optimization, and the ability to multisource capacity.

Their network allows users to keep the network hardware and software they have installed or to own or share an EDS*Net Regional Node. These nodes use off-the-shelf PBXs, multiplexers, facsimile, PADs and custom EDS software supporting up to sixteen T1s. Services include messaging, digital encrypted voice, traffic statistics, facsimile and data transmission.[31]

EDS's major customer is, naturally, General Motors. Unilever PLC has also been a large customer. EDS's target markets are financial services and banking, health care, and government. EDS acquired Anacomp's retail banking software package and is integrating it into a financial information system for BancOne Corp. of Columbus, Ohio. Terms of the contract allow EDS to resell the package elsewhere. One would expect them to have significant expertise in manufacturing because of a joint venture with Olivetti, several French acquisitions, and experience with Manufacturing Automation Protocol (MAP) being championed by GM. EDS bid with US Sprint for the $5 billion FTS-2000 government network where they are competing against AT&T/Boeing/CSC and Martin Marietta/Northern Telecom but dropped out after quibbling over responsibilities for one year.

General Electric Information Services Co. (GEISCO), Rockville, Maryland. GEISCO offers contract development work across a broad range of industries and a line of software products designed to interact with the company's information services network. Typical of GEISCO's development work is the new Applelink package that the company developed jointly with Apple Computer; after development, the package was revised to let IBM PC and Apple Macintosh users communicate via the 30-country GE Network for Information Exchange (GENIE). Users can access such features as a self-developed text database, bulletin boards, electronic mail, EDI, and the Vestor market analysis database.

GEISCO recently enhanced the international nature of its network by bringing 36 additional countries on-line through an agreement with the Dutch Group 800 N.V. GE's acquisition of RCA leaves open the option of integrating the 130-country telex network.[32]

With a proven track record in developing systems in insurance, finance, health care, retailing, and distribution and an international focus owing to GE's worldwide operations, GEISCO is an established player in the IOS market.

McDonnell-Douglas. McDonnell-Douglas has been among the most active of the defense/aircraft companies that have moved into information systems and services, although Boeing, Martin Marietta, and TRW are also visible. McDonnell-Douglas Information Systems Company was formed to hold the information systems and services acquisitions that include Microdata and

Tymnet as well as the internally developed McDonnell-Douglas Automation Company (now called McAuto). Microdata is the largest of the Big Three EDI systems suppliers, eclipsing both GEISCO and Sterling/Ordernet.[33] Tymnet holds a 40 percent market share in the packet-switching service industry and boasts ten consecutive years of profitability.

The Bell Operating Companies

The BOCs' participation in enhanced services is guided by the FCC's CI-III decision of 1986, to an even greater extent than AT&T's participation is. The decision allowed BOCs to provide enhanced services, including protocol conversion, without an arm's-length subsidiary if they implemented Open Network Architecture (ONA) plans by February 1988. Enhanced services were permitted earlier if a BOC filed intermediate plans for Comparably Efficient Interconnection (CEI) covering appropriate network elements and the CEI was approved by the FCC and the Department of Justice.

The ONA plans are designed to unbundle the BOC networks into basic service elements so that an enhanced service provider (which would be competing with the BOCs) could order precisely the services required. The plans must strive to provide competitors with "full technical equality" for service elements that are to be defined using service providers' input. Judgment of "equality" will be heavily weighted by perceptions of the alternate provider, an arrangement that appears to be something of a "guilty until proven innocent" situation for the BOCs. Network changes must be shared with service providers on a timely basis, and this information will be subject to nondisclosure agreements.

It is informative to review examples of the BOCs' ventures in enhanced services.

- Early in March 1987, Bell Atlantic became the first BOC to seek permission to provide an enhanced service under the provisions of CI-III. The service, an electronic bulletin board capability resident on a central office switch, would allow users to record messages that other callers could access with the proper code. At this time, Bell Atlantic is interested in providing a messaging capability without information content "to test service and market assumptions." CEI plans will be forthcoming.

- Although the use of "line of business" waivers is commonly associated with acquisitions by BOCs, it is not always clear whether one is or is not needed. Recent NYNEX acquisition activity provides an example of this process. NYNEX Information Resources of Lynn, Massachusetts, has been considering an investment in or acquisition of CorpTech, a provider of printed directories and information databases on high-tech companies. The need to seek Justice Department and FCC approval would depend on exactly how CorpTech's product is to be used and on the extent of NYNEX's control.
- Pacific Bell has recently teamed with Mead Data Central of Dayton, Ohio, to co-market Mead's Lexis legal database. Mead provides local bar associations with access to Lexis, and PacBell's sales force targets small law firms with an integrated service composed of the Lexis access and tariffed Centrex or data lines.[34]

Strategic Partnerships

Formal or informal joint ventures are a very common means for information services companies to develop applications, extend current products, or obtain access to vertical marketing expertise. Generally a major computer or communications company seeks to establish a beachhead in a vertical or foreign market by entering into a strategic partnership. Financial institutions, in particular, have been vigorous initiators of partnerships because of their strong in-house systems expertise, the information intensiveness of their business, and the global nature of their markets. Successful partnerships combine a long-term perspective with effective management of divergent interests.

Citicorp provides an excellent example of banking activity in information services. Citicorp's acquisition of Quotron has been compared with GM's acquisition of EDS in that it brings more systems integration talent in-house and forms a platform for Citicorp's information business. It also illustrates the risks involved with trying to supply information to competitors. Many feel that Quotron's contracts with Citicorp's competitors will be increasingly contestable as they begin to expire.

In the fall of 1986, Citicorp introduced Express Money Services, which will compete with Western Union's consumer-based money transfer system. The service builds on a service and equipment relationship with National Business Systems of Ontario. Marketed to consumers and small businesses, Express Money Services will send up to $10,000 to any of 2,000 reception points in the United States within 15 minutes. The sender can initiate a transaction in person or by telephone, with cash or a major credit card, and the service is available 24 hours a day, seven days a week.

There have also been some consolidations among Citicorp's recent activities. The Citibank Financial Account was an attempt to determine whether consumers felt comfortable banking by long-distance through strictly electronic means. Although there were some takers in the test cities—Omaha, Tampa, Minneapolis, and Atlanta—the reception was cool enough for the service to be closed down. The realization that many consumers will not generally conduct active, transaction-based banking business long-distance has caused a major shift in the evolution of Citicorp's national banking strategies. The new goal is to buy broadly dispersed, in-place branch systems and augment them with the electronic banking center technology.[35] The experiment highlights the capability that a competitor in a single vertical market has to experiment with and feed back information to the market planning and marketing strategy function within the organization.

AT&T's Lack of Competitive Advantage

We have examined the evolving conditions in the enhanced communications services markets, focusing on forecasting AT&T's ability to compete successfully as a supplier of third-party, communications-intensive IOSs.

AT&T has attempted to compete in this market in the past, with disappointing results. In 1975, the company first talked about a product that would allow dissimilar machines and networks to communicate with one another. First called Advanced Communications Service, the network was to have been a packet-switching offering, with all the intelligence for tasks like message formatting and transmission residing in AT&T switches. It was to interface with teletype machines and dumb terminals, with much of the applications software left up to the customer to provide. Failures in marketing strategy and product planning, several reorganizations, high turnover in development, and four name changes combined to make it the wrong product at the wrong time, targeted at no particular market. When Net 1000 was discontinued in June 1986, no more than twenty sales had been made.[36] Even if AT&T has learned a great deal from this experience, the cost in both financial terms and marketplace credibility makes it a most expensive way to acquire the skills we have identified as essential.

Often the real causes of failure are hidden from AT&T top management. AT&T's history as a regulated monopoly makes it even more difficult for top management to identify key problem areas. The "cradle to grave" corporate culture that existed at AT&T has two important implications for competing in enhanced services. First, in-house marketing skills of the requisite type are almost nonexistent among those senior managers who spent the formative years of their careers in a regulated monopolistic environment. Second, the market-specific product and procurement processes we have identified are difficult to target without some industry-specific expertise. AT&T has tried to address these problems by hiring marketing managers, many of whom have come from IBM, but IBM's service-intensive marketing strategy addresses only the first marketing difficulty. It is interesting to note that IBM has addressed the second area through a series of marketing agreements with independent software vendors. In May 1986, IBM signed an exclusive marketing agreement with Hogan Systems of Dallas, the largest vendor of primary systems to large banks. A similar agreement with Sterling Software for check processing software in March 1987 embodies IBM's strategy for the financial services market.[37] Yet when Citicorp acquired Quotron, AT&T decided to back out of its joint venture with Quotron to integrate PC/Unix, Starlan, and the Quotron Q1000 workstation.

We have noted the attractiveness of the enhanced services market in general and have seen the financial strength, vertical market expertise, and international experience of the players who are attracted to this market.

Perhaps least promising for AT&T's prospects is the increasingly commodity-type nature of the services in which the company currently excels. We have seen that multisourcing for these services is not only available but is in some cases preferable from both a technical and an economic perspective. And AT&T will be required to provide equally attractive communications facilities to its competitors in IOS.

Thus, AT&T enjoys no special advantage in competing in basic enhanced services where the value to be added stems from the data and processing that are provided. Likewise, vertical integration does not imply any competitive advantage in supplying third-party IOS, since an efficient market will exist for increasingly commodity-like bandwidth products.

The marked disagreement of these conclusions with AT&T's perceptions of the skills that will be needed to compete, and with AT&T's assessment of its own competitive position, is most striking. The competitive deficiencies that we have highlighted can be filled by any of the following: an internal development plan, strategic partnerships, or an acquisition of a viable competitor in value-added systems. As we have seen, internal development can be difficult and time-consuming, and it requires skilled people who command premium compensation and most frequently thrive in entrepreneurial environments. Strategic partnerships can flourish only if both partners have sufficient staying power and managerial skills to handle divergent goals objectively. Finally, AT&T must decide whether it can make a timely acquisition and then create value by acquiring a player in a market characterized by intensive human capital and labor mobility. We feel that the challenges posed by the acquisition would probably be matched by its attractiveness.

Notes

1. Frost and Sullivan, cited in *CommunicationsWeek,* April 11, 1988, p. 52.
2. AT&T, 1984 annual report, pp. 8, 11.
3. *The Wall Street Journal,* April 16, 1987.
4. "AT&T Poised to Gain Credibility," *The Wall Street Journal,* April 7, 1988, p. 6.
5. "Overall Computing Intensity," provided by the Diebold Group, cited by *Computerworld,* April 6, 1987, p. 79.
6. "Focus: VSAT Update," *Computerworld,* April 1, 1987, p. 74.
7. Peter Petre, "How to Keep Customers Happy Captives," *Fortune,* September 1985, p. 44.
8. "Texas Air Corp.'s Reservation Unit Sues American," *The Wall Street Journal,* November 24, 1987, p. 44.
9. "VA Prepares Bid," *CommunicationsWeek,* April 16, 1987, p. 12.
10. *CommunicationsWeek* and Booz, Allen & Hamilton, "ISDN—Putting

the Vision in Focus," special section of *CommunicationsWeek,* December 15, 1986, p. 22.

11. Like the basic rate, the EDSL's B channels are 64 kbps; however, at the primary rate the D channel is also 64 kbps. Total usable bandwidth is thus 1.536 Mbps (in North America). Depending on the configuration, the primary rate D channel can be used for signaling or packet data, or it may be used as another B channel, with channel control assigned to another facility's D channel.

There are four interfaces defining the choices a user has for accessing the ISDN-based network; the R, S, T, and U interfaces. Essentially they are different points corresponding to how closely the customer premise equipment matches the protocol and bandwidth of the ISDN channel being accessed.

12. *CommunicationsWeek,* December 15, 1987, p. 13.

13. Ibid., p. 10.

14. Ibid., p. 11.

15. Ibid., p. 23.

16. Personal communication with senior Bell Canada officer.

17. *CommunicationsWeek,* January 26, 1987, p. 1.

18. Eric K. Clemons and Steven O. Kimbrough, "Information Systems, Telecommunications, and Their Effects on Industrial Organization," *Proceedings of the Seventh International Conference on Information Systems,* December 1986.

19. "Texas Air Corp.'s Reservation Unit Sues American," *The Wall Street Journal,* November 24, 1987, p. 44.

20. W. Schatz, "On the Edge," *Datamation,* April 15, 1986.

21. With current technology (565 Mbps, 1 and 3 protection scheme), this amounts to 72,500 working channels at 64 kbps. Since 1 for 3 protection involves the provision of a fourth idle channel for each three working channels installed, this represents a further reservoir of potential excess bandwidth.

22. *CommunicationsWeek,* January 26, 1987, p. 7.

23. This requirement comes in the form of a ruling that the company must modify its network to provide competitors with open network architecture (ONA).

24. IEEE Institute, news supplement to *IEEE Spectrum,* April 1987.

25. Company presentation, University of Pennsylvania, The Wharton School, Philadelphia, Penn., November 19, 1986.

26. Charles L. Howe, "Another Tangled Network," *Datamation,* March 15, 1986, p. 64.

27. Philip Kotler, *Marketing Management: Analysis, Planning and Control* (Englewood Cliffs, N.J.: Prentice-Hall, 1980), p. 251.

28. F. Warren McFarlan and James L. McKenney, *Corporate Information Systems Management: The Issues Facing Senior Executives* (Homewood, Ill.: Richard D. Irwin, 1983), p. 184.

29. *CommunicationsWeek,* April 13, 1987, p. 28.

30. Ibid., same date and March 9, 1987, p. 6.

31. *CommunicationsWeek,* February 2, 1987, p. 20.

32. Ibid., December 22, 1986, p. 17.

33. W. Schatz, "On the Edge," *Datamation,* April 15, 1986.

34. *Datamation,* April 13, 1987, p. 37.

35. *American Banker,* March 27, 1987, p. 1.

36. Charles L. Howe, "Another Tangled Network," *Datamation,* March 15, 1986, p. 64.

37. *American Banker,* March 2, 1987, p. 22.

10

The Hidden Barrier to the Bell Operating Companies and Their Regional Holding Companies' Competitive Strategies

Richard L. Nolan, C. Rudy Puryear, and Dan H. Elron

The hidden barrier to the competitive strategies of the Bell operating companies (BOCs) and their regional holding companies (RHCs) is computer technology. This may come as a surprise, in that their parent company, AT&T, built the most advanced telephone system in the world, and for the lowest cost. Furthermore, AT&T spawned the most innovative and productive high-tech research facility in the world: Bell Labs has historically produced a patent a day, and many of these patents have provided underlying technologies for major companies in the United States (e.g., IBM) and abroad (e.g., Sony, Philips). Nevertheless, it is our contention that although some of their chief executive officers may not fully realize it yet, the divested BOCs are low-tech companies competing in a high-tech industry. We think that transforming themselves into high-tech businesses is their major strategic challenge.

The purpose of this chapter is to support our proposition that the major barrier to transformation of the BOCs is their current position in information technology. To accomplish our purpose, we first present a framework for business transformation and then, using this framework, position the RHCs. Next, we outline

Notes: We use the phrase *information technology* to denote the broader role of computers in processing data in office automation, in factory automation, and in modern communications. This broader role is in contrast to a traditional narrower role in data processing.

the role and position of information technology for the RHCs and their BOCs. Finally, we conclude with a definition of the information technology issue.

The Need for Transformation

Deregulation and divestiture have caused the RHCs to search for new business strategies and organizational structures to fulfill their new visions. Because the old vision and structure served so well for so long, the process has not been easy.

Fifty years ago, Theodore Newton Vail led the Bell System to its greatness with the vision of "universal service": a telephone for every house in the United States. The capital investment of laying copper wire was tremendous—so tremendous that Vail steered the Bell System toward a monopoly structure. The monopoly structure was granted on the basis of the need to protect the investor's capital by guaranteeing a market and a price for universal telephone service. In addition, Vail was instrumental in establishing a framework that enabled the vision of universal service to become a reality. First, a two-tier pricing structure was created for personal phone service and business phone service. Since the value to business was assumed to be higher, businesses were charged a higher rate than households; effectively, business subsidized home service. Second, long-distance service was also charged a premium, based on the assumption that long-distance service was a luxury. Third, although a return-on-equity basis was decided upon as fair for pricing telephone service, the depreciation life of equipment was extended (often to forty years) to keep the price of telephone service as low as possible.

For more than fifty years, the Bell System operated as the model monopoly. By 1984, it was the largest company in the world, with revenues exceeding $100 billion. But by then the electronic computer had fundamentally changed the telecommunications industry. Computers were now a key component of the telephone network, replacing the old electromechanical switches that had been used for more than five decades; today's telephone switches are essentially computers, with more than two-thirds of their development cost attributable to software. The computers in the telephone network, with their enormous flexibility and ever increasing functionality, revolutionized the traditional network, creating a wealth of new markets and opportunities. Almost suddenly, the fairly stable telephone business became as volatile as the computer business.

Changes in telecommunications technology were accompanied by the rapidly increasing importance of communications. As the world moves toward a "global economy,"[1] telecommunications have become a major factor in gaining competitive advantage and improving productivity. Sophisticated airline reservation systems (such as those used by American Airlines, United Airlines, and British Airways) and distributed manufacturing systems with electronic linkages to suppliers (such as those used by General Motors, Ford, Fiat, Boeing, and McDonnell-Douglas) are just two examples of the vital role that telecommunications play in business.

The increasing importance of telecommunications in the marketplace and the new technological opportunities attracted a variety of new entrants to the telecommunications business. The Bell System was perceived as being too slow to adapt to new technology. In short order, on January 1, 1984, it was broken into eight parts: one long-distance carrier—AT&T—and seven RHCs, encompassing the twenty-two BOCs and providing local telephone service within some 160 Local Access and Transport Areas (LATAs). Figure 10-1 shows the breakup of AT&T into seven RHCs and summarizes some of the key structural changes that occurred after AT&T's divestiture.

Today, these seven RHCs are viable, $8 to $12 billion dollar businesses, providing local telephone services to households and businesses in the various states. Currently, all of them are doing well in returning profits to their stockholders.

But fundamentally the RHCs' basic business of voice communications, which still represents more than 80 percent of telephone network traffic, is a slow-growth business. In many cases it is not expected to grow at more than 5 percent per year. In fact, Ameritech's number of customer lines increased by less than 6 percent between 1982 and 1986. This situation presents a major challenge for the RHCs, which have promised their shareholders significant earnings growth rates. Already, NYNEX has announced that it expects earnings growth to slow in 1987.[2] A senior executive at one RHC summarized the situation very well

Figure 10-1
AT&T and Twenty-two Bell Operating Companies to Seven Regional Holding Companies and Twenty-two Operating Companies

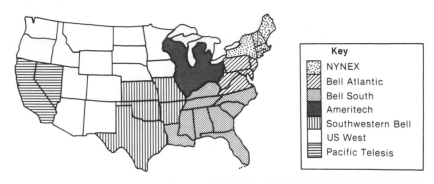

Key
- NYNEX
- Bell Atlantic
- Bell South
- Ameritech
- Southwestern Bell
- US West
- Pacific Telesis

	AT&T (pre-1984)	Regional Holding Companies (post-1984)
Business	• Telecommunications	• "Information services"
Business Strategy	• Universal service	• Being developed
Structure	• One company; fully regulated monopoly	• Seven independent regional holding companies, partially regulated
Products	• POTS (plain old telephone service) for residential service. Business service, long-distance service	• Multiple and diverse customer-driven services for both residential and business applications
Organization	• Centralized research (Bell Labs) • Centralized planning • Centralized product development • Centralized manufacturing (Western Electric) • Decentralized operations and service • Centralized information systems development and maintenance • Decentralized computer operations	• RHC-level research, contracted research and centralized research (Bellcore) • RHC/BOC planning • Decentralized product development (by each RHC) • Contracted manufacturing (with Western Electric and competitors) • BOC-level operations and service • Centralized and increasingly RHC-specific information systems development and maintenance • BOC-level computer operations

Figure 10-2
The Time-Bomb Chart

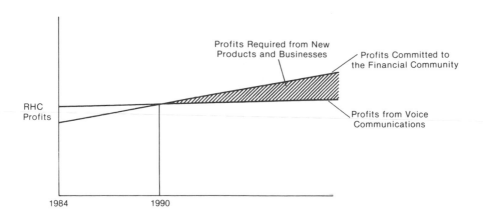

with what he referred to as a "time-bomb" chart. (See Figure 10-2.)

In addition to being a slow-growth business, the RHCs' business is also one with rapidly changing technology but obsolete regulatory policies, such as a two-tier pricing structure, luxury premium pricing for long-distance service, and forty-year depreciation schedules for equipment that is obsolescing in an eighth of the time. It is a business in which Vail's sixty-year-old vision of universal service has been achieved and in which a new vision, for the twenty-first century, is needed.

The Business Transformation Framework

In our consulting work, we use a framework shown in Figure 10-3, for describing business transformation.

The pathway begins with the creation of a consensus among the executives that transformation is required—that is, that the current strategy is not the right one. This step involves a realistic assessment of the company's existing and future environment and of its current strengths and weaknesses. After a consensus is reached, the executives can be mobilized to develop a

Figure 10-3
Business Transformation "Pathway" Framework

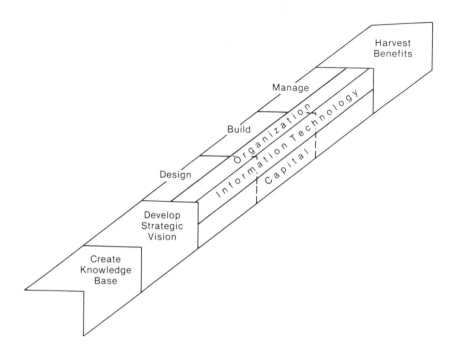

new strategic vision. From that strategic vision, organization, information technology, and capital structures can be designed, built, and implemented. Finally, the last step in the pathway is installing a harvest process to ensure that the benefits of the new business strategy will be realized in real terms.

RHCs and Transformation

We believe that the RHCs have negotiated the first two steps of the transformation framework to varying extents. After the breakup of the Bell System, no one would deny that the RHCs needed to develop a new vision. The merging of telecommunications and computing and the advent of the "information revolution" have led virtually all to the same conclusion concerning the business that they should be in. This point is succinctly stated in the NYNEX 1986 annual report: "At NYNEX, we're in the Information Business."

Due to the common starting point of the RHCs, it is perhaps not surprising that their CEOs' strategies seem to have several similarities. In keeping with the traditionally conservative approach of the Bell System, these executives aim at building on their "principal business of telecommunications" through cost control and modernization of the network. However, they are simultaneously venturing into new markets "to become leading companies in the Information Industry."[3]

As part of that strategy, several of the RHCs are trying to become "systems integrators"—providing telecommunications network linkage to their large clients' diverse, geographically dispersed computer systems, ranging from mainframe computers, word processors, and individual workstations to devices like automated teller machines (ATMs) or robots used in manufacturing. This approach is a modern version of the old Bell System goal of providing complete, end-to-end telecommunications service. This technologically ambitious strategy is not unique: companies such as IBM, AT&T, Boeing, EDS, and GE have targeted this business as well.

The RHC diversification strategy has been quite aggressive. While still restricted from providing long-distance service and manufacturing telephone equipment in the United States, the RHCs have entered a variety of businesses, often focusing on the "information industry" and outside the regulation umbrella.

Southwestern Bell has acquired Metromedia's nationwide cellular radio and paging operations for $1.2 billion. Southwestern has also expanded its directory-publishing business outside its region. NYNEX purchased IBM's product center stores. Ameritech has acquired Applied Data Research (ADR), a leading database software firm. (ADR was sold in 1988 at a significant loss.) NYNEX is arguing that since a majority of international long-distance traffic coming into the United States comes through New York, NYNEX should be able to enter the international market. Indeed, the RHCs are also venturing into the overseas manufacturing of customer premise equipment (CPE) and the sale of CPE in the United States and abroad.

The RHCs seem to expect that continuing deregulation will eventually allow them to integrate their new businesses with the traditional network. With continuing pressure on their part to deregulate,[4] it is indeed conceivable that the RHCs will soon have even fewer constraints to developing wide-ranging competitive strategies for becoming "leading companies in the Information Industry."

In positioning the RHCs on our business transformation pathway, we note that all seem to have reached the point of establishing strategic visions. The next step in the pathway is the design of the organizational, information technology, and capital structures necessary to implement the strategic visions. Although we will leave the assessment of the organizational and capital structures to those more qualified to judge, our cursory analysis leads us to believe that those structures will not be the major constraints for the RHCs.

In regard to organization, deregulation seems to be proceeding steadily. Clearly, what is commonly called corporate culture is a constraint in moving from regulated, engineering-driven companies to unregulated, market-driven companies. But this is not a hidden constraint.

Figure 10-4
The Stages of Computing Growth

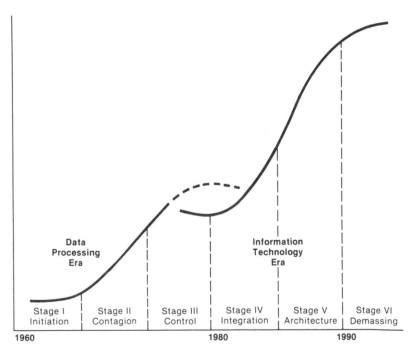

Source: Adapted from Richard L. Nolan, "Managing the Crises in Data Processing," *Harvard Business Review* (March–April 1979), Exhibit 1, pg. 117.

In regard to capital, significant efforts to cut costs, including staff reductions, have allowed the RHCs to consistently improve their earnings per share since divestiture. Most RHC stocks nearly doubled from 1983 to 1986, possibly making it feasible at this time to raise capital through a stock offering.

We think, however, that information technology is the blind-side constraint. Using a stages framework[5] for describing the evolution and maturity of information technology in companies, we can position RHCs' business visions as requiring advanced-stage capabilities but not having reached advanced-stage information technology maturity.

Figure 10-4 divides the historical evolution of computers in companies into roughly two eras. The first era, termed the *Data Processing Era*, generally initiated in the late 1950s, early 1960s and lasted about twenty years. This is the era in which the computer was applied to transaction processing. The second era, the *Information Technology Era*, is built on the capabilities established in the first era and adds diverse distributed information technology, some of which involves levels of integration to give the company a competitive edge or strategic advantage.

Current RHC Strategies Require Advanced-Stage Systems Capabilities

For many years the Bell System was extremely successful at pursuing its vision of providing universal service. Over time, it developed an organizational structure and operational procedures that allowed it to very effectively deliver its major product—basic telephone service—to millions of customers throughout the United States. One of the keys to the Bell System's success was the fact that essentially the same service was provided to all its customers, allowing the same procedures to be used everywhere.

This situation changed dramatically, however, with the advent of the transistor and the information revolution. Whereas twenty years ago a telephone company customer had little choice beyond basic telephone service—referred to then as POTS (plain old telephone service)—today technology has made a variety of products and options available. The BOCs offer advanced new products for voice traffic as well as for the data processed by the computers that thousands of their customers now own. These new offerings are often very advanced and are increasing at a dramatic pace. They include "custom features," such as call forwarding; electronic banking; automated alarm services; and dedicated, high-speed data networks. The new integrated services digital network (ISDN) technology[6] will allow the integration of voice and data, which is expected to lead to an even more advanced generation of products within a few years.

The RHCs are finding that there is increasing demand for these new products. They are also discovering, however, that their market is no longer homogenous. Advanced features have fragmented the market for basic telephone service, while the business customers' different computing environments and business strategies, such as just-in-time manufacturing, are creating a multitude of customer requirements. In addition, cost has become an issue: business customers, who may be spending hundreds of millions of dollars a year on telecommunications, are increasingly price-sensitive in this era of global competition; now that deregulation has made it feasible, they do not hesitate to switch suppliers when cost savings can be found.

Serving this suddenly fragmented, competitive marketplace is a major challenge for the RHCs. The history of AT&T's divestiture illustrates that new telecommunications markets attract a variety of new competitors. The RHCs have decided to encourage that competition; a major element of their competitive strategies is the push for deregulation.

As part of the drive for deregulation, several RHCs are endorsing the concept of open network architecture (ONA), which would allow access by RHC competitors to any part of the network. The RHCs believe that if competition is allowed, they will be permitted to sell a variety of new products and features that they are barred from selling today. This step would support the RHCs' goal of ensuring that those new products be provided to their customers through their telephone network, rather than by the customers' purchasing their own equipment.

Yet in order to take advantage of the current and future opportunities, the RHCs must undergo a major transformation: they must change from being a provider of a limited set of products in a noncompetitive environment to being a provider of low-cost, individualized customer services, such as systems integration.

We believe that this transformation is analogous to that observed in the manufacturing industry, in which technology has allowed companies to dramatically increase the range of the products they manufacture: through computer-driven flexible manufacturing systems, different products can be produced based on customer demand, with little or no retooling costs.

We can use the automotive industry as an example. Here the evolution started with the traditional assembly line, which produced thousands of virtually identical cars (leading to Henry Ford's famous statement that you could have a car of any color, so long as it was black). This was the hard-wired manufacturing plant, where any product change required major retooling. In the telephone industry, retooling is analogous to the telephone line rewiring in the switch that was required to change or add telephone service.

In the automotive industry, technological advances such as computers and robots have allowed an increase in flexibility. Today, robots on the assembly line install such options as larger engines or different seats, depending on the configuration with which they are programmed. In the telephone industry, the advent of the electronic switch has allowed customers to choose such options as call waiting and call forwarding. Some business customers are already reconfiguring their BOC-provided Centrex telephone exchange, using a personal computer, with no manual intervention by telephone company employees.

As technological advances increase the potential for automation and flexibility, the manufacturing process in the automotive industry will become increasingly automated. At some point, it is predicted that we will move to "lights-out" manufacturing—a factory with no employees. This kind of manufacturing can already be seen in the machine parts industry.[7]

The telephone companies are already talking about the lights-out manufacturing environment, which would allow them to provide low-cost customized service. Some have termed this concept *Instant Service*. Instant Service involves the automation of the customer service order process from the time the order is placed to the availability of dial tone at the customer's new address. Today, that process is one of the most complex, cumbersome, and error-prone processes in the BOCs.

Under the concept of Instant Service, the service order flow shown in Figure 10-5 would be completely automated and integrated so that the customer could receive service immediately after submitting an order.

Although the economic feasibility and value of providing im-

Figure 10-5
The Instant Service Process Flow

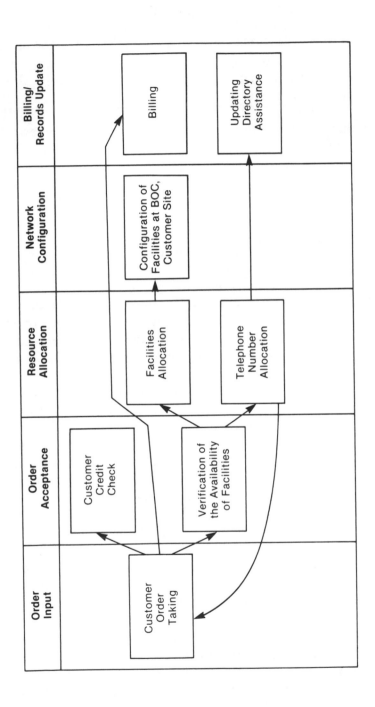

Figure 10-6

The Transformation of the Telephone Network from a Hard-Wired
Manufacturing Plant to Lights-Out Manufacturing

	Mass Production ⟶ Customization		
	Hard-wired Manufacturing Plant	**Flexible Manufacturing**	**Lights-Out Manufacturing**
Telephone network	Electromechanical switches	Electronic/digital switches	Fully automated network
Network characteristics	Inflexible; changes require major effort	Some programming of features, functions	Completely flexible; all changes can be programmed
Service provisioning	Telephone workers needed to change line assignment in order to provide new service	Some service features can be programmed automatically	New service requires no manual intervention
Number of products	Very limited	Some variety	Fully customized
Focus of management efforts	High efficiency	Increased flexibility	Satisfy unique customer needs
Major information technology requirements	Efficient automation of routine tasks (e.g., billing)	Access to data; system integration	Fully integrated technology architecture (including order entry systems, network computers, and administrative systems)

mediate service to most customers are debated in the industry, the concept exemplifies the telephone companies' objective of achieving maximum efficiency and responsiveness through automation and their version of lights-out manufacturing. Also, just as in manufacturing, the concept of Instant Service has a dramatic impact on the RHCs' information technologies. Figure 10-6 shows the broadening of the business strategy and the corresponding new requirements for information technology.

Computer systems are no longer used solely in the traditional support roles of the Data Processing Era, when their main purpose was to automate clerical or transaction-processing tasks such as billing and record keeping. Instead, computer systems have become a key component in providing new products that the RHCs sell, in customizing the service provided to each customer, and in automating complex, labor-intensive processes. Various computer systems, including the telephone switches, must be integrated to support processes like Instant Service. They must also be designed to be flexible and efficient, in order to respond to market needs.

The Current RHC Systems' Base Is at an Early Stage of Maturity

The Bell System, with its emphasis on efficiency and technology, has always invested heavily in automation. As the world's largest corporation, it had both the resources and capabilities to do so. This investment has resulted in the BOCs having a massive systems infrastructure. The computer systems traditionally used by the BOCs can be classified into four broad categories:

1. *Administrative systems.* These traditional data processing systems were the first to be developed. As in most companies, their purpose was to automate billing and accounting activities, which were extremely labor-intensive. Many were developed to support the financial reporting requirements of regulators.

2. *Service support systems.* These systems deal with the process of providing and tracking customer service, and they include the systems used to allocate lines and telephone numbers and to provide directory assistance information. The service order systems usually feed the billing systems with information about a new customer's service.

3. *Network support systems.* These systems keep track of network components, such as telephone trunks and switching equipment, and help control some of the network functions. They are often based on dedicated minicomputers and include very sophisticated diagnosis and repair systems.

4. *Network systems.* These systems include the switches and telephone equipment that route telephone calls, record usage, and provide features such as call waiting and call forwarding.

Figure 10-7
Computer Systems Used by the BOCs

System Category	Typical Functions Supported	Technology Base	System Development
Administrative systems	• Accounting • Billing • Budgeting	• Mostly large mainframe computers	• Mostly BOC developed; some developed by AT&T
Service support systems	• Service order entry • Directory assistance • Tracking of service locations • Allocation of facilities	• Mostly large mainframe computers • Some minicomputers	• Mostly BOCs (some other vendors)
Network support systems	• Inventory tracking • Network tracking, repair	• Mainframes for inventory management • Minicomputers for network control, repair	• Mostly AT&T
Network computers	• Call routing • Call recording • Advanced features	• Dedicated computers embedded in the switches, other parts of the network	• Mostly AT&T

In Figure 10-7 we have summarized, for each system category, the functions supported, the technology base, and the systems development origin. These systems represent an extremely large investment. A number of the BOCs are spending between 9 and 12 percent of their revenues on just the first three system categories, and may have more than 5,000 employees operating and supporting the systems. The investment may amount to more than $1 billion dollars a year (in expenses and capital expenditures) for a typical RHC. In contrast to other segments of the telephone business, which are cutting back on staff and expenditures, this is an area in which hiring and growth are still occurring.

The BOCs are some of the most heavily automated businesses in the country, often operating extremely large and complex systems.

> **Example:** TIRKS* (Trunk Integrated Record-Keeping System) is one of the world's largest systems, whose purpose is to store information about only part of a telephone company's facilities and to assist in planning and engineering circuits. It has more than two hundred databases and requires enormous computing capacity; more than two hundred people currently support and update it.

The Bell System's early start in automation, coupled with its massive investment in computer systems, would appear to be a major competitive advantage for the RHCs in their new strategies. But this is not the case; on the contrary, their information technology is more of a liability than an asset with respect to their desired transformations.

Many of the systems now operated by the BOCs were developed more than ten years ago; some were built in the early sixties. Since then, much of the technology has changed. More important, the science of building reliable, maintainable systems through software engineering is maturing only today.[8] Many of these systems are thus obsolete; they are often large and difficult to maintain. In addition, most BOCs have a "Tower of Babel" of different technical environments, with a multitude of computer languages and operating systems; in some cases, new computer languages and operating systems were specially developed to support a new system development project.

> **Example:** One BOC has more than twelve types of minicomputers and mainframe computers, with more than twenty different operating systems. The diversity of technologies often occurred because systems were developed separately for different business areas; integration was not a major concern. In some cases, diversity was intentional: AT&T did not want to depend too heavily on one vendor and did not want to be perceived as favoring a specific vendor. For the systems developed in the BOCs, diversity was also caused by the BOCs' choice of vendors in their service area: Northwestern Bell chose Honeywell for its billing system, while Michigan Bell elected a Burroughs environment.

*TIRKS is a registered trademark of Bell Communications Research, Inc.

Maintaining such a variety of systems and adapting them to new business requirements have become a major challenge for the BOCs. Going beyond the support required for day-to-day activities and integrating the existing systems into the four systems categories, as required for Instant Service, are often impossible.

Already, systems are placing significant burdens on the essential functions of today's business. This is particularly true when important data are unnecessarily duplicated.

> **Example:** One BOC has more than six customer databases, each supposed to contain the same basic information. Currently, more than two hundred employees are required to ensure that the information agrees across systems.
>
> In another BOC, a senior executive estimated that more than five hundred staff members were employed to reconcile unnecessarily duplicated systems.

In addition to the costs involved, duplication of data has an impact on management decision making. In several cases, differences and errors in data stored in different locations lead managers to reach different conclusions based on which database they happen to use, resulting in wasted time and frustration. Although this situation often occurs in other companies, for the BOCs it creates a significant disadvantage: the BOCs usually have considerably larger, older, and more heterogeneous systems and have not yet learned to resolve the problems they are encountering now that they are beginning to make important, market-based decisions. In addition, those problems are accentuated at the RHC level, where several BOCs are involved. For example, Ameritech has five different billing and customer record systems—one for each BOC (Indiana Bell, Illinois Bell, Michigan Bell, Ohio Bell, and Wisconsin Bell). Similar situations occur in the other RHCs, making it virtually impossible to track the total relationship between an RHC and a customer (such as IBM) who purchases products from every BOC.

The systems infrastructure is also affecting the BOCs' efforts to introduce new products.

> **Example:** A BOC marketing department introduced a set of new features, to be sold as part of a trial. It found that it would have to wait two years before the existing billing system could be

modified to accommodate the new product. The product manager threatened to hire dozens of temporary employees (in a period of layoffs) to manually send out bills. When a concern was raised that customers would react unfavorably to receiving two bills, the whole trial was discontinued.

In another BOC, a new type of data service was introduced. All billing for this service is performed manually; it is estimated that only half the bills sent out are correct.

These examples show some of the ways in which technical issues have had a major impact on how computer systems support the business. This impact is demonstrated further in Figure 10-8, which describes a sample of the RHCs' major systems in terms of two attributes:

- Technical quality: The quality of their design and their ease of maintenance and operation.
- Functional quality: The quality of support that the systems provide to their users.

The information presented in Figure 10-8 is based on our analysis of more than one hundred key systems in BOCs belonging to multiple RHCs between 1984 and 1986. The figure is divided into four quadrants, based on our standards for systems of acceptable quality. These benchmarks were empirically derived, based on the analysis of several thousand systems in several hundred large companies in the United States and abroad. The benchmarks are represented by the dotted lines in the diagram. Systems in the top-right quadrant are thus functionally and technically acceptable, while systems in the lower-left quadrant are not performing satisfactorily in either respect. Systems in the lower-right quadrant are technically acceptable but have low functional quality; this may indicate that they were not designed to suit the users' needs. Systems in the upper-left quadrant are of acceptable functional quality but are of poor technical quality; this results in high maintenance costs and, in the long run, is also likely to lead to a deterioration of their functional quality, since it is difficult to adapt such systems to changing business requirements.

As we can see in the figure, the generally more recent, AT&T-developed network support systems—some of which are now supported by Bellcore, the joint RHC support organization—are of relatively better technical quality. Probably because their role

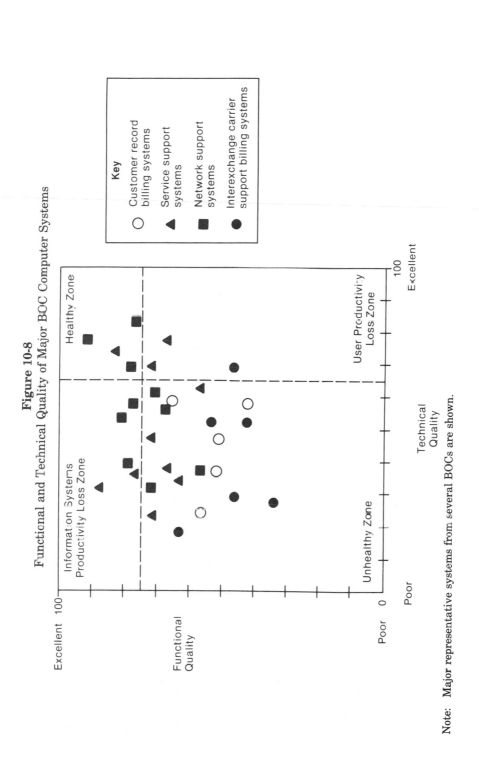

Figure 10-8

Functional and Technical Quality of Major BOC Computer Systems

Key

○ Customer record billing systems

◀ Service support systems

■ Network support systems

● Interexchange carrier support billing systems

Healthy Zone

Information Systems Productivity Loss Zone

User Productivity Loss Zone

Unhealthy Zone

Functional Quality

Excellent 100

Poor 0

Technical Quality

Poor

Excellent 100

Note: Major representative systems from several BOCs are shown.

has not been heavily affected by divestiture, they are generally considered functionally satisfactory.

The service support systems are usually of lower functional quality, often because they cannot accommodate the users' new requirements and new products. The BOCs are investing heavily in this area in their drive toward Instant Service-related goals: increased efficiency and reduction of errors. Several of the systems being installed in this area are very advanced; in some cases, a large proportion of the manual activities involved in providing service have been eliminated. However, a major constraint the BOCs face in effectively implementing these systems is the need to integrate them with the often much older billing and customer record systems. The latter systems are not satisfying the BOCs' needs, and they are extremely difficult to maintain and adapt to business and regulatory changes.

An interesting case is that of the carrier billing and support systems. These systems had to be rapidly developed prior to divestiture in order to allow the BOCs to bill their new customers—AT&T and other interexchange carriers—for access to the local network. Because of the short development time, unclear business requirements, and lack of large systems development experience in the BOCs, these systems are often of very poor technical and functional quality. In many cases they consume a large part of the BOCs' maintenance resources.

We believe that the low functional quality of most of the systems noted in Figure 10-8 is not due to poor design at the time they were developed. In fact, many of the systems represented the state-of-the-art at that time; they still provide functionality and reliability rarely found among many *Fortune 500* companies. However, they have been difficult to adapt to new requirements and, in particular, to the new postdivestiture business environment. Because the BOCs rely so much on automation, they are particularly affected by any shortfalls in their systems that have an immediate impact on their business activities. To a certain extent, today's BOCs are thus prisoners of the Bell System's heavy investment in computing early in the Data Processing Era.

Some of the requirements that the BOCs' systems in each of the four categories need to address are described in Figure 10-9.

Two key issues emerge from an analysis of Figure 10-9. First, all system categories need to support new marketing-related

Figure 10-9
Requirements for the Major BOC Systems

System Category	Major Systems Requirements	
Administrative systems	• Provide billing for new products •Support new postdivestiture customers—the long-distance carriers • Provide information on operating performance and costs in support of business decisions rather than for regulatory reporting	
Service Support Systems	• Streamline the service provisioning process • Reduce manual intervention; assist the sales process • Easily adapt to support new products	• Integrate systems to provide Instant Service • Support requirements for management information
Network support systems	• Support direct customer control of network features • Eliminate manual intervention • Support the "self-healing, software-driven" network concept • Support new products	
Network computers	• Eliminate manual activities • Support new products, features • Support deregulation, provide ONA capabilities	

needs, including new products and the needs for management information about customers, markets, and product usage. This is a relatively new area of emphasis for the BOCs, whose systems traditionally focused on automating basic operational tasks such as billing and inventory tracking. A major realignment of system priorities to support the new market-driven strategies is required.

Second, integration of systems is a major new challenge. It is needed to support strategies such as Instant Service and to satisfy customers' needs for a variety of advanced, individualized products. The BOCs' current difficulties in tailoring, tracking, or billing for new products present an opportunity for smaller and nimbler competitors that use flexible, state-of-the-art systems to target specific market niches. These difficulties could become a major concern for the RHCs when ONA allows competitors to access the BOCs' equipment directly.

Can the RHCs Overcome Their Information Technology Constraint?

In evaluating the capability of a company to rapidly improve its systems infrastructure and evolve to advanced stages, we examine the status and condition of three areas that we believe are critical to that process: information technology resources, management, and users.

Information Technology Resources

Before divestiture, the BOCs depended heavily on AT&T for the development of major systems, particularly network support systems such as TIRKS. Although the BOCs operated the systems they received from AT&T, they did not learn how those systems were built or how to maintain them. The BOCs have become excellent at operating systems; we have found that they have some of the most efficient computer centers in the United States. However, they have not developed the staff and the skills to design and manage large development projects in the service and network support categories.

> **Example:** In one BOC, 60 percent of systems development personnel were found to be below the skills level that their management considered appropriate for the postdivestiture era.

Example: Many BOCs are embarking on new development projects. However, they have trouble finding experienced project managers and resort to using junior employees, thereby significantly increasing the risk of project failure.

The same applies to telephone network-related activities. Most BOC projects relating to the telephone network do not involve much more than adding capacity, often achieved through adding plug-in modules to central offices. When the more advanced switches in the BOCs' central offices have needed reprogramming, the task has been traditionally carried out by the manufacturer—in many cases AT&T, a company the BOCs now regard as a competitor.

The BOCs are relying on vendors for some of their most strategic activities, including their experiments with the latest ISDN technology. They do not have any other choice. In addition, the BOCs suffer from a lack of up-to-date technical knowledge, partly because they have done little hiring in recent years, and partly because the marketplace for telecommunications experts is very competitive. A senior BOC executive, commenting on his RHC's strategy of becoming "a leader in the information technology field," said that his was "a low-tech company in a high-tech industry." The BOCs do not have the technical R&D specialists that Bell Labs had.

The fact that the BOCs lack some critical technology functions to support their basic business, let alone to support their new strategies, was anticipated at the time of divestiture. The organization that was established to take AT&T's place and perform R&D, as well as product and systems development, for the BOCs was Bell Communications Research—Bellcore. Bellcore, which is jointly owned by the seven RHCs, does indeed perform a critical role in supporting the RHCs' technology needs. It develops standards and plans for the telephone network and maintains and enhances AT&T-developed systems such as TIRKS. However, we believe that it cannot be considered a long-term solution to the technology requirements of the RHCs' competitive strategies.

Although the RHCs realize that they need Bellcore's technical expertise, major problems have begun to surface. Whereas AT&T owned the BOCs, Bellcore is owned by the RHCs. Key decisions on what Bellcore should do are now made jointly by seven owners, who often have divergent needs and strategies and consider one another competitors. One of the major RHC concerns is that, as AT&T is perceived to have done, Bellcore will not satisfy each RHC's individual needs. Many in the RHCs consider AT&T's "centrally developed systems" to be poorly suited to their needs; an example often cited is that systems installed in the rural Midwest incorporate features needed only in the New York region.

Bellcore has tried to address some of the RHCs' concerns by beginning to develop projects for individual RHCs, working as a contractor. However, this arrangement has not yet proved effective. US West has threatened to pull out of Bellcore altogether, in part because it is concerned about the potential disclosure of its highly proprietary network strategy to the remaining RHCs.[9]

At this point, it appears that Bellcore will remain critical in planning for the future of the telephone network and in supporting major network-related systems. However, Bellcore may not be as well positioned to help the BOCs to integrate these systems with their existing infrastructure, particularly their homegrown administrative systems. Because of this, and because of the competition among the RHCs, the RHCs will probably not be able to rely solely on Bellcore in the development of particularly aggressive technology-based strategies.

The Users

Users of information technology in the RHCs have traditionally been accustomed to receiving new products and systems as they were released by AT&T, with little consultation. Now they find themselves in the position of having to specify their needs and identify new opportunities for using technology. Many do not have the knowledge and experience required and do not know how to work together with information technology specialists on developing new projects.

Even the users who have developed some computer skills, possibly outside the RHCs, are rapidly growing frustrated with the particularly complex and inflexible current BOC computing infrastructure. Often, they purchase their own personal computers so that they can develop solutions independently.

However, the users realize very quickly that the key to their problems lies in using data stored in the major systems' databases, such as data about the purchase patterns of telephone service features by specific market segments. In many cases they find that information difficult or impossible to access. For functions such as strategic planning and market and product management, this situation creates a major barrier to employee effectiveness and productivity.

The Management Structures and Controls

Managing their own destiny with respect to information technology is also a new challenge for the BOCs. Information technology was rarely an issue at the level of the BOC officers. Prior to divestiture, it was rarely considered a strategic issue, and many of the required decisions were made by AT&T. This approach has resulted in a lack of the management structures required to direct and control information technology activities.

> **Example:** One BOC was planning the allocation of funds to computing, using it the same way as it planned for network expansion projects, only on capital budget required for each yearly budget period. This method has resulted in projects being reconsidered every year, in dysfunctional start-stop funding, and in few completed projects.

The lack of independent development activities within the BOCs has also resulted in limited development of the controls, procedures, and methodologies required to develop complex computer systems. In examining a variety of areas, ranging from human resource planning and recruitment strategies for information technology specialists to software acquisition procedures and auditing of software projects, we found several BOCs that had the level of controls found in large companies eight to ten years ago.

A particular concern was the lack of data management procedures. Without such procedures, current activities in systems development will continue to result in duplicated data and in situations wherein data is difficult to access by the users.

In analyzing some of the technology management practices at the BOCs, we found that BOC managers, often experts in telephone technology, believe that the management of computing is essentially similar. This is a major fallacy: managing data processing and developing systems is very different from operating a hard-wired switching network.

This finding illustrates a fundamental point: because a company operates its technology successfully does not mean that it can perform all the functions required to develop and manage it.

Are External Sources an Answer to the Information Technology Constraint?

The RHCs are beginning to understand the extent of their requirements for technological expertise. Several of them have turned to the outside. In addition to Ameritech's purchase of ADR, US West purchased a vendor of banking software, and NYNEX acquired several software firms.

But these acquisitions do not seem to be the answer. Many of the purchased firms are too small or too specialized to address the issues faced by their RHCs. In addition, there is a risk that by focusing the acquisitions on their internal issues, the acquired firms will lose their distinctive market position. Purchase of a larger company, such as EDS, could potentially solve this problem. However, not many firms of that size are available for purchase, and the challenges of integrating them with the RHCs' environment and using them effectively could be significant—as demonstrated by the recent GM/EDS merger. The difficulties of integrating the new acquisitions were demonstrated when Ameritech decided to sell ADR in 1988, after expected synergies with the rest of the company did not materialize.

Conclusions

The computer has changed the telecommunications industry dramatically. The strategies that made AT&T the largest and one of the most successful companies of the twentieth century will no longer work. AT&T's seven RHC offspring need a new set of strategies.

They are searching for these strategies, and the answer seems to lie in becoming major players in the information industry. While all seven RHCs seem to realize this point, the information technology infrastructure that they inherited at divestiture appears to be more of a liability than an asset to their aggressive strategies; it represents a major constraint in their attempts to pursue these strategies. We have concluded that four major challenges must be addressed by the RHCs' top management to change their current information technology liability into a strategic asset:

1. Renovation. The current major administrative systems need to be renovated and reengineered to meet new business requirements.
2. Integration. The administrative, service, and network systems need to be re-architected into an integrated, responsive system of product delivery.
3. Flexibility. A high degree of flexibility (similar to that of a flexible manufacturing system) will need to be designed into the overall technical infrastructure of the administrative, service, and network systems to accommodate new products, technologies, and regulatory changes.
4. Management information. The marketplace is increasingly fragmented, volatile, and demanding. Accordingly, management information systems need to be developed to determine marketplace demands, customer responsiveness, and profitability. Timely management information will be critical to gaining market share and to maintaining satisfactory profit margins in the new competitive environment.

Until these very significant challenges are met, the RHCs will be unable to realize their full potential in the highly competitive information industry marketplace.

Notes

1. Alvin Toffler, *The Third Wave* (New York: Bantam Books, 1981), pp. 168–193.

2. *The Wall Street Journal*, May 7, 1987.

3. NYNEX Corporation, 1986 annual report, p. 6; Southwestern Bell Corporation, 1986 annual report, p. 4; and Bell Atlantic, 1986 annual report, p. 3.

4. "The Baby Bells Take Giant Steps," *Business Week*, December 2, 1985, p. 94.

5. Richard L. Nolan, "Managing the Crises in Data Processing," *Harvard Business Review* (March–April 1979): 115–126.

6. Paul Wallich and Glenn Zorpette, "The Information Revolution Awaits," *IEEE Spectrum* (November 1985): 104.

7. Ramchandran Jaikumar, "Postindustrial Manufacturing," *Harvard Business Review* (November–December 1986): 69–76.

8. Richard L. Nolan and Gonzalo Verdugo, "Renovating the DP Applications Portfolio," *Stage by Stage* (Nolan, Norton & Co. publication) (May–June 1986): 1–11.

9. "US West to Drop Bellcore," *MIS Week*, January 12, 1987, pp. 1, 6.

ABOUT THE CONTRIBUTORS

Joseph S. Baylock
Joseph S. Baylock is a consultant with the telecommunications practice of Gartner Group Inc., where he specializes in competitive analysis, carrier strategies, and international and satellite communications. He received a B.S. in computer engineering from Rensselaer Polytechnic Institute and an M.B.A. from the Wharton School at the University of Pennsylvania, where he made his contribution to this publication. While in industry, Mr. Baylock held positions in marketing, finance, and operations with several carriers and manufacturers.

Stephen P. Bradley
Stephen P. Bradley is the chairman of the Managerial Economics area at the Harvard Business School and currently teaches Industry and Competitive Analysis in the M.B.A. program. Professor Bradley's research interests center on competitive strategy and particularly strategy analysis models. He is the author of numerous articles and three books, *Management of Bank Portfolios* (1976), *Applied Mathematical Programming* (1977), and *Quantitative Methods in Management* (1978). He received a B.E. in electrical engineering from Yale University in 1963 and an M.S. and Ph.D. in operations research from the University of California at Berkeley in 1965 and 1968, respectively.

Eric K. Clemons
Eric K. Clemons is associate professor of Decision Sciences at the Wharton School and of Computer and Information Science at the Moore School of Electrical Engineering, both at the University of Pennsylvania. His research and teaching interests include strategic information systems, telecommunications technology and its commercial application, decision support systems, and the use of information systems in international securities trading. Dr. Clemons is project director for the Reginald H. Jones Center's Sponsored Research Project in Information Systems, Telecommunications, and Business Strategy. His education includes an S.B. in physics from MIT and an M.S. and Ph.D. in operations research from Cornell University. Dr. Clemons is currently a member of the National Council of the Association for Computing Machinery and associate editor of *Management Information Systems Quarterly*.

Dan H. Elron
Dan H. Elron is a senior manager in the New York office of the consulting firm of Nolan, Norton & Co., and a member of its telecommunications industry practice. He has consulted with many organizations in the telecommunications industry in the United States and abroad, including most of the regional Bell Operating Companies. His work has focused primarily on the development of long-term strategies and information technology architectures. He has specialized in addressing the issues arising out of the extensive need for information sharing in the regional Bell Operating Companies and the need to integrate traditional information systems and the telephone network to support new customer-driven strategies.

T. Grandon Gill

T. Grandon Gill is a doctoral student in Management of Information Systems at the Harvard Business School. His concentrations include artificial intelligence, expert systems, and advanced telecommunications technologies such as VSAT and fiber optics. Prior to entering the doctoral program, he was a communications officer in the Navy and vice president of Technical Services at Agribusiness Associates, a specialized strategic consulting firm in Wellesley Hills, Massachusetts. He received his M.B.A. from Harvard in 1982.

Jerry A. Hausman

Jerry A. Hausman is professor of Economics at MIT, where he teaches courses in telecommunications, econometrics, microeconomics, and public finance. He has conducted extensive research and written many articles in all three areas. He received his A.B. from Brown, and his M.Phil. and D.Phil. from Oxford, where he was a Marshall Scholar. He is a fellow of The Econometric Society; he was awarded the 1980 Frisch Medal from that society; in 1985, he was awarded the John Bates Clark Award by the American Economic Association. Mr. Hausman also consults in the areas of telecommunications, energy economics, and antitrust—he is currently a member of the Governor of Massachusetts' Advisory Council for Revenue and Taxation and the Committee for National Statistics.

Peter W. Huber

Peter W. Huber is a lawyer and writer, specializing in telecommunications policy and liability law. He has a Ph.D. in mechanical engineering from MIT, where he taught for six years. He received his law degree from Harvard and clerked for Judge Ruth Ginsburg on the D.C. Circuit Court of Appeals and then for Justice Sandra Day O'Connor of the U.S. Supreme Court. He is the author of *Liability: The Legal Revolution and Its Consequences* (Basic Books, 1988), *The Geodesic Network* (U.S. Department of Justice, 1987), as well as numerous papers and articles, ranging from scholarly publications in law reviews, to magazine pieces for *The New Republic,* to newspaper articles in *The Wall Street Journal.* He has appeared on "Face the Nation" and "The McNeil-Lehrer News Hour." Huber prepared the first triennial report of the Department of Justice on the break-up of AT&T.

Elon Kohlberg

Elon Kohlberg is a professor at the Harvard Business School. He received a B.Sc. (1966), M.Sc. (1967), and Ph.D. (1973) in mathematics from the Hebrew University in Jerusalem. His main research interests are competition and game theory, on which he has published extensively. His teaching responsibilities have included courses on Competitive Decision Making, Game Theory, and Industry and Competitive Analysis. He has also pioneered a course on Competition in the Telecommunications Industry.

F. Warren McFarlan

F. Warren McFarlan is professor of business administration and director of research at the Harvard Business School, where he has been a member of the faculty since 1964. His most recent book, with James I. Cash and James L. McKenney, is *Corporate Information Systems Management: The Issues Facing Senior Executives* (Irwin, 1988). His current area of research is the use of information systems technology for competitive advantage.

James L. McKenney

James L. McKenney is professor of business administration at the Harvard Business School, where he has been a member of the faculty since 1960. His most recent book, with James I. Cash and F. Warren McFarlan, is *Corporate Information Systems Management: The Issues Facing Senior Executives* (Irwin, 1988). His current areas of research are the management of corporate communications and the impact of electronic systems on management work.

Richard L. Nolan

Richard L. Nolan is chairman and cofounder of Nolan, Norton & Co., a leading international organization of counselors to management, focusing on the effective management of computer-based technologies. Dr. Nolan is responsible for the strategy and direction-setting activities of the firm. The author of seven books and more than 100 published articles on data processing management, he has consulted with dozens of organizations worldwide, including Du Pont, IBM, and Digital, as well as several large telecommunications providers. He is the originator of the Stages Theory for analyzing data processing growth, a theory he researched and developed while an associate professor at Harvard Business School. His current research focuses on how organizations are transforming themselves to the twenty-first century, and on the key role that information technology plays in that process.

Roger G. Noll

Roger G. Noll is professor of economics at Stanford University and director of Stanford's Public Policy Program. He is also a faculty member in political science at Stanford's Graduate School of Business. Professor Noll is the author of five books and approximately 100 articles on a variety of topics in economics, political science, and public policy, including several studies of aspects of communications policy. He is the co-author, with Linda R. Cohen, of a forthcoming book, *The Technology Pork Barrel,* dealing with the economics and politics of government programs to develop new technology for private industry. His current research includes studies of state regulation of telephones, the evolution and policy implications of administrative processes in environmental regulation, and the development of regulatory institutions in Midwestern farm states in the 1870s.

H. Edward Nyce

H. Edward Nyce, executive vice president, Manufacturers Hanover Trust Company, is officer-in-charge of MH Information Technology Services and is responsible for the corporation's competitive utilization of technology. He is a graduate of Temple. From 1957 to 1963 he was an account representative with IBM; in 1963 he joined Reigel Paper, where he served as director of Management Information Services; and in 1971 he joined Pentamation Enterprises as a division president. He moved to Manufacturers Hanover in 1973 and was named to his present position in 1986.

James P. O'Neill

The late James P. O'Neill was a CPA who was employed by the Ford Motor Company from 1951 to 1968—his last assignment as vice president finance for Ford Europe. From 1968 through 1982 he worked for Xerox in several capacities, including president of Business Products Group, president of Information Technology, and executive vice president and chief staff officer. He was also a member of the Xerox board of directors. At the time of his death, he was with Regional Financial Enterprises.

Bruce M. Owen

Bruce M. Owen is president of Economists Incorporated, a Washington, D.C. consulting firm specializing in antitrust and regulatory issues. He is also an adjunct professor of Public Policy at Duke University. Mr. Owen was formerly the chief economist of the Antitrust Division of the U.S. Department of Justice, and, earlier, of the White House Office of Telecommunications Policy. He was graduated with a B.A. in economics from Williams College in 1965; later he received a Ph.D. from, and taught at, Stanford University. Mr. Owen is the author or co-author of numerous articles and several books, including *Television Economics* (1974), *Economics and Freedom of Expression* (1975), *The Regulation Game* (1978), and *The Political Economy of Deregulation* (1983).

C. Rudy Puryear

C. Rudy Puryear is a managing principal in the consulting firm of Nolan, Norton & Co., and is responsible for its telecommunications practice and the overall marketing efforts of the firm. He has consulted with a variety of major organizations, including several of the regional Bell Operating Companies, the Bank of America, R. J. Reynolds, Royal Trust of Canada, Lockheed, and Du Pont. In the telecommunications industry, he has assisted the senior management of several large companies in evaluating their post-divestiture information technology position and in developing comprehensive plans for the strategic use of information technology.

Kevin R. Sullivan

Kevin R. Sullivan is a partner with Pillsbury, Madison & Sutro in Washington, D.C., specializing in communications and litigation. After graduating from the law school of Catholic University in 1975, Mr. Sullivan was an Honor Program appointee to the Antitrust Division of the Department of Justice. He became assistant chief of the Communications and Finance section and lead counsel responsible for the Modified Final Judgment in the AT&T case between 1984 and 1986. He had supervisory responsibilities in domestic and international communications and financial services for the division, and served as liaison for national security and emergency preparedness issues as they related to telecommunications.

Richard H.K. Vietor

Richard H.K. Vietor is a professor at the Harvard Business School, where he teaches courses on the regulation of business and the international political economy. His research, which focuses on business-government relations, has been published in numerous journals and several books; among the most recent, are *Energy Policy in America Since 1945* (1984), *Telecommunications in Transition* (1986), and *Strategic Management in the Regulatory Environment* (1988). Before coming to Harvard in 1978, Professor Vietor taught at the University of Missouri. He earned a B.A. from Union College, an M.A. from Hofstra University, and the Ph.D. in history from the University of Pittsburgh (1975).

Index

Note: Antitrust cases in italics.